Revision

NEW GCSE SCIENCE

Physics

for AQA A Higher

Author: Nicky Thomas

Revision guide +
Exam practice workbook

William Collins' dream of knowledge for all began with the publication of his first book in 1819. A self-educated mill worker, he not only enriched millions of lives, but also founded a flourishing publishing house. Today, staying true to this spirit, Collins books are packed with inspiration, innovation and practical expertise. They place you at the centre of a world of possibility and give you exactly what you need to explore it.

Collins. Freedom to teach

Published by Collins
An imprint of HarperCollinsPublishers
1 London Bridge Street
London
SE1 9GF

Browse the complete Collins catalogue at
www.collins.co.uk

British Library Cataloguing in Publication Data
A Catalogue record for this publication is available from the British Library.

Project managed by Hart McLeod Limited, Cambridge.

Edited, proofread, indexed and designed by Hart McLeod Limited, Cambridge.

Printed in China

Acknowledgements
The Authors and Publishers are grateful to the following for permission to reproduce photographs.

p11 ©Shutterstock
p17 ©Alan Williams / Alamy
p32 fig 2 ©Shutterstock
p32 fig 3 ©Shutterstock
p33 ©Shutterstock

Whilst every effort has been made to trace the copyright holders, in cases where this has been unsuccessful, or if any have inadvertently been overlooked, the Publishers would gladly receive any information enabling them to rectify any error or omission at the first opportunity.

About this book

This book covers the content you will need to revise for GCSE Chemistry AQA A Higher. It is designed to help you get the best grade in your GCSE Chemistry Higher Exam.

The content exactly matches the topics you will be studying for your examinations. The book is divided into two major parts: **Revision guide** and **Workbook**.

Begin by revising a topic in the Revision guide section, then test yourself by answering the exam-style questions for that topic in the Workbook section.

Workbook answers are provided in a detachable section at the end of the book.

Revision guide

The Revision guide (pages 6–48) summarises the content of the exam specification and acts as a memory jogger. The material is divided into grades. There is a question (**Improve your grade**) on each page that will help you to check your progress. Typical answers to these questions and examiner's comments, are provided at the end of the Revision guide section (pages 49–53) for you to compare with your responses. This will help you to improve your answers in the future.

At the end of each module, you will find a **Summary** page. This highlights some important facts from each module.

Workbook

The Workbook (pages 66–108) allows you to work at your own pace on some typical exam-style questions. You will find that the actual GCSE questions are more likely to test knowledge and understanding across topics. However, the aim of the Revision guide and Workbook is to guide you through each topic so that you can identify your areas of strength and weakness.

The Workbook also contains example questions that require longer answers (**Extended response questions**). You will find one question that is similar to these in each section of your written exam papers. The quality of your written communication will be assessed when you answer these questions in the exam, so practise writing longer answers, using sentences. The **Answers** to all the questions in the Workbook are detachable for flexible practice and can be found on pages 113–125.

At the end of the Workbook there is a series of **Revision checklists** that you can use to tick off the topics when you are confident about them and understand certain key ideas.

Additional features

Throughout the Revision Guide there are **Exam tips** to give additional exam advice, **Remember boxes** pick out key facts and a series of **How Science Works** features, all to aid your revision.

The **Glossary** allows quick reference to the definitions of scientific terms covered in the Revision guide.

Contents

Energy

Storing and transferring energy

- Everything stores energy. The more energy things store, the more work they can do. The amount of energy stored is measured in Joules (J).

- Energy cannot appear or disappear if anything changes, but some energy may spread from the energy stores to the surroundings.

Energy and chemistry

- Energy is not a physical substance. In a chemical reaction, the mass of chemicals before and after is the same, but the amount of energy stored in the chemicals changes. Some energy spreads to the surroundings, or may be absorbed from the surroundings.

Remember!
We detect energy being transferred or stored in different places in different ways. However, energy cannot exist in different forms.

Figure 1: Examples of things that transfer energy

Infrared radiation

Emission absorption and uses

- All objects emit and absorb infrared radiation.

- Shiny light coloured surfaces absorb infrared radiation slower than black matt surfaces. They reflect infrared radiation well.

- Hotter surfaces emit radiation faster than cooler surfaces.

- We can design objects to reduce the rate of energy transfer.

Remember!
Objects are absorbing and emitting infrared radiation at the same time. If infrared radiation is emitted faster than it is absorbed, then the object cools down.

How Science Works

- Black surfaces inside an oven emit infrared radiation better than shiny surfaces. This helps heat to transfer to the food cooking inside the oven.

Figure 2: Comparing the emission and absorption of infrared radiation by different surfaces

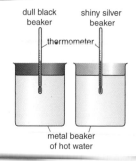

Infrared radiation and global warming

- Many things affect the rate of global warming. One theory is that sea water is dark coloured and will absorb infrared radiation from the Sun faster than ice does.

Remember!
Objects warm up if they absorb more infrared radiation than they emit.

Improve your grade

Explain whether a kettle of hot water cools down quicker if its outer surface is coloured white or dark green.
AO2 (3 marks)

Kinetic theory

Bonds between particles

- Bonds between particles are strongest in solids and weakest in gases.

- Melting is when a solid changes to a liquid. When the solid is heated, particles vibrate more vigorously. Bonds between particles break and reform so particles can change places.

- Freezing is when a liquid cools and changes to a solid.

- Boiling is when a liquid changes to a gas when the liquid is heated. The particles break their bonds and can move around randomly.

- Condensing is when a gas cools and changes to a liquid.

D–C

What's a particle?

- In solids, particles vibrate and are held together by strong bonds.

- In liquids, particles move around each other but cannot escape due to weak bonds.

- In gases, particles move randomly as the bonds are very weak.

Remember!
The particles do not change as they change states, but they behave differently.

EXAM TIP
Remember to use the terms particles and bonds in answers to questions about changes of state.

B–A*

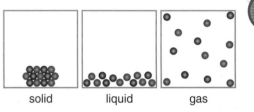

Figure 3: Arrangement of particles in each state of matter

solid   liquid   gas

Conduction and convection

Transferring the energy

- If one end of a solid is heated, particles vibrate more and shake their neighbours. Conduction transfers energy from one particle to another through the solid.

- If one part of a liquid or gas is warmed, the particles vibrate more taking up more space. The warm region of gas or liquid expands, becoming less dense and rising above cooler denser regions. This is convection.

- Convection currents spread heat through liquids or gases that are heated from the base, or cooled from the top.

Figure 4: The dye traces the convection current as the water warms

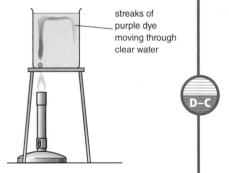

streaks of purple dye moving through clear water

D–C

Why are metals good conductors?

- Electrons in metals are free to move rapidly from hotter to cooler regions, transferring energy.

B–A*

 Improve your grade

Explain why convection can take place in a liquid but not in a solid. **AO2 (3 marks)**

Evaporation and condensation

Evaporation and environment

- Evaporation is when a liquid changes to a gas at temperatures lower than the boiling point.

- Evaporation is quickest if it is warm (more particles have enough energy to break bonds linking them), if the liquid has a large surface area, or if it is windy (so vapour above the liquid's surface does not become saturated).

- Water evaporates from the Earth's surface, and cools forming clouds. Water vapour condenses into liquid rain. This is the water cycle.

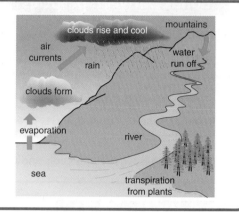

Figure 1: The water cycle

D–C

Evaporation and pressure

B–A*

- At high altitudes, wind speeds are high and air pressure is low. It is easier for particles to evaporate.

Remember!
Evaporation occurs at the surface at low temperatures, but boiling occurs throughout the liquid at boiling point.

Rate of energy transfer

Conduction, infrared radiation and convection

D–C

- When hot objects cool, energy is transferred to the surface by conduction and from the surface by convection. Hotter objects lose energy more quickly than cooler objects.

- The ratio of surface area to volume affects the rate of heat loss. If the surface area is larger, energy is transferred quicker from the object.

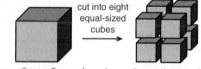

2 cm x 2 cm x 2 cm cube
volume = 8 cm³
surface area = 24 cm²

1 cm x 1 cm x 1 cm each cube
volume = 1 cm³
surface area = 6 cm²
total volume of the eight cubes = 8 cm³
total surface area of the eight cubes = 48 cm²

Figure 2: Cutting the block increases its surface area

Cooling curves

B–A*

- The cooling curve shows the temperature drops more quickly at hotter temperatures.

Figure 3: A typical cooling curve for water. Room temperature is 20 degrees C.

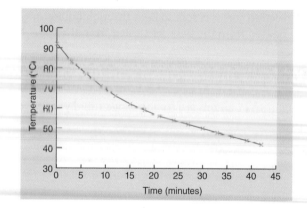

Improve your grade

Explain why a towel dries quicker on a windy summer day. **AO2 (3 marks)**

Insulating buildings

How much can we save

- Payback time is the time taken to save as much money as it costs to install an energy saving measure.

- Payback time in years = installation cost ÷ annual saving

Method of insulation	Installation cost (£)	Annual saving (£)	Payback time (years)
loft insulation	240	60	4
cavity wall insulation	360	60	6
draught proofing doors and windows	45	15	3
double glazing	2400	30	80

Figure 4: Payback time for different energy saving measures

The wider picture

- The energy needed to manufacture and install insulation should be less than the energy saved by using it. Some councils give grants to install different types of insulation.

Specific heat capacity

Calculating absorbed energy

- Energy is transferred to an object that heats up, or from an object that cools down.

- Energy transferred to or from a material (J) = mass (kg) × specific heat capacity (J/Kg °C) × temperature change (°C)

- For example, the specific heat capacity of water is 4200 J/Kg. The energy absorbed by 250 g of water heated from 20 °C to 100 °C is 0.25 × 4200 × 80 = 84 000 J

Oil filled radiators

- Oil filled radiators use oil with a specific heat capacity of 2000 J/Kg °C. They reach a higher temperature than water filled radiators for the same energy input.

Remember!
You must always use the temperature difference in specific heat capacity calculations.

 Improve your grade

The specific heat capacity of copper is 390 J/Kg °C. Explain whether copper heats up quicker than the same mass of water when they are put in a hot place. The specific heat capacity of water is 4200 J/Kg °C. **AO2 (3 marks)**

Wasted energy

- The Law of Conservation of Energy says energy cannot be created or destroyed when it is transferred. All energy is usefully transferred, dissipated or stored. The energy that spreads to the surroundings is called wasted energy.

- A Sankey diagram shows energy transfers in a device.

- The energy input is shown at the left side of the arrow.

- The arrow splits. Each section shows the output energy form.

- The width of each part of the arrow shows the proportion of energy it represents.

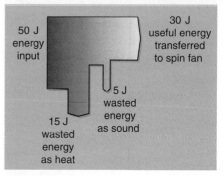

Figure 1: Sankey diagram for an electric fan

Comparing devices

- Sankey diagrams compare the proportion of energy transferred usefully by different devices.

Remember!

All energy must be accounted for during a transfer. It may be stored, usefully transferred or dissipated to the surroundings.

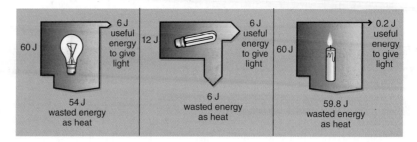

Figure 2: Sankey diagram comparing different light sources

Efficiency

Efficiency and wasted energy

- Efficient devices do not waste much energy. Efficiency = useful energy out ÷ total energy in. The answer is always a decimal less than 1.

- Convert decimals to percentages by multiplying by 100. Efficiency is always less than 100%.

- You can use information from a Sankey diagram to calculate efficiency.

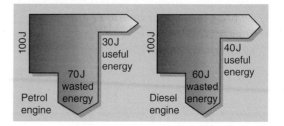

Figure 3: These diagrams show the petrol engine's efficiency is 30% and the diesel engine's efficiency is 40%

Perpetual motion machines

- Perpetual motion machines cannot be made as they would be 100% efficient, which is impossible.

Improve your grade

Explain what this Sankey diagram shows in as much detail as possible.
AO2 (3 marks)

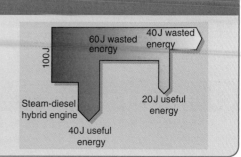

Figure 4:

Electrical appliances

Alternatives to electricity

- Electrical appliances transfer the energy supplied by electricity into something useful.

- Mains electricity is convenient, safe and pollution free at the point of use. However, generating electricity in power stations causes pollution.

- Batteries change chemical energy into electricity when the battery is part of a circuit.

- Alternatives to electricity include gas or biomass for cooking, lighting and transport.

- Some appliances work using solar power or energy stored in springs instead of batteries.

LEDs

- LEDs produce light but work in a different way to light bulbs. Electrons in materials used to make LEDs absorb energy from a supplied voltage. Then they release this energy as light. The light is a single colour because the same energy is absorbed and released by all electrons in an LED.

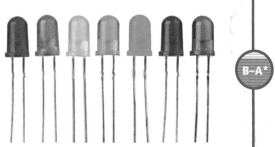

Figure 5: The colour of light from LEDs depends on the energy released by the electrons inside it

Energy and appliances

How much energy is used?

- Power is the rate that something does work, or the rate of energy transfer. It is measured in Watts (W) or kilowatts (kW).

- Calculate the energy used by an appliance in joules using: power (W) × time (s)

Heat and current

- Larger currents carry more energy than smaller currents. More powerful equipment uses thicker cables than less powerful equipment.

- Connecting cables are designed to heat up as little as possible when they carry a current so that as little energy is transferred to the surroundings as possible.

- Other wires (for example, in heating or lighting filaments) are designed to become very hot when a current flows through them so they can transfer energy to the surroundings as light or heat.

> ### EXAM TIP
> Check you know the right units for time, power and energy. You will lose marks if you write **j** instead of **J** for example, or if you leave them out.

Improve your grade

Two different bulbs are switched on for 10 minutes. Calculate the energy transferred by each one.
i) a 60 W filament bulb
ii) a 10 W energy efficient bulb **(AO2 4 marks)**

The cost of electricity

How much does it cost?

- A kilowatt-hour is calculated using power (in kW) × time (in hours).
  - For example, a 0.1 kW light bulb switched on for 15 hours uses 1.5 kWh (0.1 kW × 15 h).
- The cost of using electricity is the number of kilowatt-hours used × cost per kilowatt-hour.
  - For example, if each kilowatt hour costs 12p, the cost of using the light bulb was 18p (1.5 kWh × 12p/kWh).

Figure 1: A typical electricity bill

Switching to standby

- Leaving equipment on standby for long periods of time increases our electricity usage.

Power stations

Energy changes

- In a power station, fossil fuels (coal, gas, oil) are burned, or nuclear fuels (uranium, plutonium) undergo fission. Water is heated, changing to steam. Steam drives turbines, which spin generators, generating electricity.
- Many power stations are about 35% efficient. Gas power stations are 60% efficient, using burning gases and steam to spin turbines. Power stations using waste energy directly for heating are 70-80% efficient.

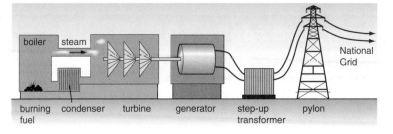

Figure 2: How a power station generates electricity

How Science Works

- Using more efficient power stations reduces environmental damage because they use less fuel and produce fewer emissions for every kWh of electricity generated.

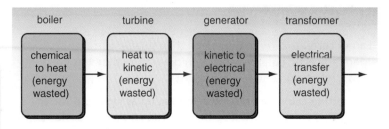

Figure 3: Energy changes taking place in a power station

What should we burn?

- Decisions on how to use inaccessible fuels must be made in future. There are advantages and disadvantages.

> **Remember!**
> Fossil fuel and nuclear power stations generate electricity the same way, but use different energy sources to provide heat.

Improve your grade

Describe the energy changes taking place in these parts of a coal fired power station: the burning fuel; the boiler; the turbine; the generator. **AO1 (4 marks)**

Renewable energy

Four more renewable resources

- Renewable energy resources will not run out. These resources spin turbines directly:
  - When the wind blows, it spins blades on a wind turbine.
  - Water trapped behind hydroelectric dams falls through pipes, driving turbines.
  - Tidal barrages are walls built across river mouths. When the tide goes in or out, water flows though pipes in the barrage driving turbines.
  - Waves drive turbines in small wave generators.
- Steam drives turbines in biomass and geothermal power stations:
  - Biomass is organic waste that is burned.
  - Heat from rocks deep underground changes water to steam in geothermal power stations.
- Photovoltaic cells change sunlight to electricity directly. This is solar power.

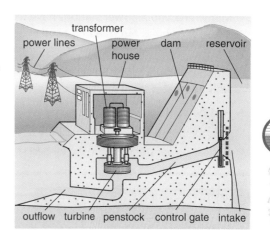

Figure 4: Hydroelectric power station

Storing energy

- Some renewable energy resources are weather dependent and unreliable. It may be possible to store some generated electricity in large underground batteries for later use.

Remember!
Renewable energy resources will not run out.

Electricity and the environment

More effects of generating electricity

Type of energy source	Advantages	Disadvantages
Fossil fuels	Does not occupy large land areas.	Produces polluting gases and solid waste. Mining damages the environment.
Nuclear fuels	Does not produce greenhouse gases.	Creates radioactive waste that must be safely stored.
Hydroelectricity	Renewable.	Causes large scale flooding. Produces greenhouse gases.
Biomass	Renewable. Reduces need for landfill sites.	Produces greenhouse gases.
Wind power	Renewable. Does not produce polluting gases.	Can affect wildlife. Must be sited in windy areas.
Tidal power	Renewable. Does not produce polluting gases.	Causes large scale flooding and changes water flow.
Geothermal	Renewable.	Can release toxic gases from below the Earth's surface.
Solar	Renewable. Does not produce polluting gases.	Large scale use involves large land areas and can affect wildlife.

Not so green hydroelectric power

- Hydroelectricity can cause more global warming than fossil fuel power stations as vegetation rots at the bottom of the reservoir and at its edges when water levels change.

EXAM TIP

If you are asked about the environmental impact of an energy source, your answer must describe environmental damage (for example, flooding, disrupting river flow, emission of greenhouse gases) and not a general disadvantage (for example expensive or unreliable).

Improve your grade

Explain whether a hydroelectric power station or a coal-fired power station is best for a city located near the coast.
AO3 (5 marks)

Making comparisons

Costs and reliability

- Many renewable power stations have high capital costs and low running costs. Fossil fuel power stations are cheaper to build but operating costs will increase as the supplies of fossil fuels fall.

- Fossil fuel and nuclear fuels can be stored, increasing their reliability. A single power station generates large amounts of electricity. Solar cells, wave and wind turbines are weather-dependent so less reliable, and each unit generates small amounts of electricity.

- Hydroelectric power stations generate electricity quickly when needed but the reservoir levels must be maintained, sometimes by pumping water back to the reservoir.

- Nuclear power stations have the longest start up and shut down times, followed by coal then gas power stations.

Electricity from sewage

- Scientists are constantly looking for new ways to generate electricity.

- Experimental methods to generate electricity include oxidising sewage using bacteria. This creates charged particles which can be separated, setting up a voltage and allowing a current to flow.

How Science Works

- You should be able to compare the benefits and drawbacks of different energy sources in different situations.

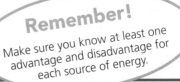

Remember!
Make sure you know at least one advantage and disadvantage for each source of energy.

The National Grid

High voltages

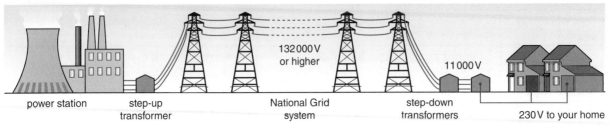

power station   step-up transformer   132 000 V or higher   National Grid system   11 000 V   step-down transformers   230 V to your home

Figure 1: The National Grid

- Electricity is transmitted from power stations to homes and businesses through the National Grid.

- Power (watts) = voltage (volt) × current (amp).

- Step up transformers at power stations increase the voltage to 400 000 V.

- Increasing the voltage in power cables reduces the current. Wires heat up less, so thinner cables are needed and less energy is wasted.

- Step down transformers in substations reduce the voltage. Electricity is supplied at 230 V in homes.

Choosing cables

- The National Grid uses aluminium cables reinforced with steel, which conduct electricity well.

Remember!
You should be able to explain what each part of the National Grid does.

Improve your grade

Explain whether a coal-firepower station or a hydroelectric power station is best to cope with surges in demand during the day. **AO3 (3 marks)**

What are waves?

Transferring energy

- Waves transfer energy from a source without transferring matter.

- Longitudinal waves oscillate in the same direction that the energy travels, and include sound waves.

- Transverse waves oscillate at right angles to the direction the energy travels, and include electromagnetic waves, water waves.

Figure 2: What is the direction of energy transfer for each of these waves?

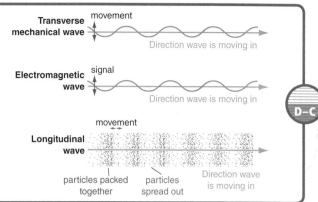

Earthquakes

- Seismic waves are mechanical waves that transfer energy from earthquakes through the earth.

- Primary waves are longitudinal waves. They travel quicker than secondary waves, which are transverse waves.

Remember!
A wave only transfers energy. The substance the wave travels through does not travel with it.

Changing direction

Refraction

- The normal is an imaginary line drawn at right angles to a surface.

- Waves can be reflected (bounce off a surface). The angle between a reflected ray and the normal is the same as the angle between an incoming ray and the normal.

- Waves can be diffracted (spread through a gap or round an obstacle). Diffraction is greatest when the wavelength is about the same size as the gap or obstacle.

- Waves can be refracted (change direction at a boundary). Waves refract because they change speed in different materials. If a wave moves into a material where it travels slower, the wave refracts towards the normal.

Figure 4: Diffraction

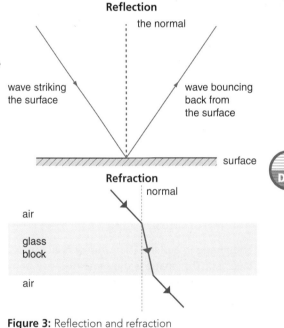

Figure 3: Reflection and refraction

Rainbows

- Raindrops refract sunlight. Blue light refracts more than red light. This is why we can see a rainbow if the Sun is behind us, shining into a rain storm.

Remember!
Make sure you know what each of the key terms means.

Improve your grade

An echo is a reflected sound wave. Explain why you can hear echoes only in certain places, and why you may hear more than one echo. **AO3 (3 marks)**

Sound

Sound is a wave

D–C

- Vibrating objects create sound waves. Sound waves are longitudinal mechanical waves and cannot pass through a vacuum.

- All sound waves can be reflected (echoes), refracted or diffracted. Loud sound waves have a large amplitude; quiet sound waves have a small amplitude.

- Frequency measures the number of cycles per second and is measured in Hertz (Hz). Humans hear sounds between 20 Hz and 20 000 Hz.

- High pitched notes have a short wavelength and high frequency; low pitched notes have a long wavelength and low frequency. The higher the frequency of a wave, the more energy it can carry.

Using an oscilloscope to compare sounds

B–A*

- An oscilloscope shows a sound wave as a trace on a screen.
  - Amplitude is the height of the trace measured from the mid point to the highest (or lowest) point.
  - The wavelength is the distance from one peak to the next peak.
  - A grid is marked on the screen.
    The time base shows how many seconds each horizontal square represents so the frequency can be calculated.

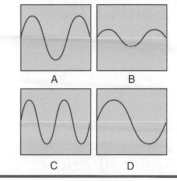

Figure 1: Sound B is quieter than sound A; sound C is higher pitched than sound D

Light and mirrors

Flat mirrors

D–C

- Ray diagrams show how one or more rays of light travel when they are reflected or refracted.

- At least two rays are needed to show where the image of the object will be seen and its appearance.

- Compared to the object, images can be;
  - upright or inverted (upside down)
  - magnified (bigger), diminished (smaller) or the same size
  - laterally inverted (left and right sides reversed)
  - real or virtual (formed where there is no light from the object).

The image in a plane (flat) mirror is virtual, upright and laterally inverted. The image is the same distance behind the mirror as the object is in front of the mirror.

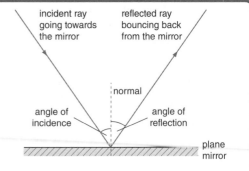

Figure 2: A ray diagram showing a reflected ray

Using flat mirrors

B–A*

- A flat mirror forms an image the same distance behind the mirror as the object is in front of it.

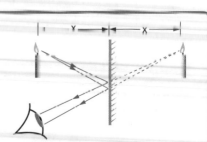

Figure 3: How an image is formed in a flat mirror

Improve your grade

Draw traces to show two sound waves. One sound is lower pitched and twice as loud as the other sound. Label the amplitude and wavelength on each trace. **AO1 (4 marks)**

Using waves

More about radiation and its uses

- Radio waves, microwaves, infrared and visible light are electromagnetic waves used in communication systems.

- Radio waves are used for radio and TV broadcasts. Local and national radio broadcasts use long wavelength radio waves. TV broadcasts use short wavelength radio waves.

- Microwaves are used in mobile phone networks and satellite communication, as the atmosphere does not absorb microwaves.

- Infrared radiation comes from anything warm. It is used in remote controls.

- Visible light allows us to see and is used in photography.

Diffraction

- Diffraction affects how well electromagnetic waves are received. If waves cannot diffract (spread) around an obstacle, the signal they carry is not received clearly.

- TV and VHF (very high frequency) signals have wavelengths of several metres which is much smaller than hills or buildings. The waves cannot diffract around them. TV and VHF receivers must be in direct line-of-sight with the transmitter.

- Long-wave radio signals have wavelengths of several kilometres, which enables them to diffract around hills and buildings. The signals are received even where shorter waves cannot be received well.

Figure 4: The signal from the transmitter comes in a straight line to these aerials

The electromagnetic spectrum

Energy and wavelength

- Electromagnetic radiation is a continuous spectrum of transverse waves carrying energy.

- In a vacuum, all electromagnetic waves travel at the same speed, the speed of light. The speed (or velocity) of the wave is calculated using:
  - speed (m/s) = frequency (Hz) × wavelength (m)

- The shorter the wavelength of an electromagnetic wave, the higher its frequency.

- Electromagnetic waves with a short wavelength and high frequency carry most energy.

- Electromagnetic wavelengths range from about 10^{-15} m (gamma rays) to about 10^4 m (radio waves).

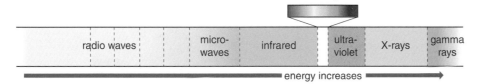

Figure 5: The electromagnetic spectrum

Electromagnetic waves

- Electromagnetic waves can be reflected, refracted and diffracted, just like waves.

Improve your grade

Calculate the speed of radio waves with a wavelength of 10 000 m and frequency of 30 000 Hz. **AO2 (3 marks)**

Dangers of electromagnetic radiation

Protecting from the dangers of radiation

- Microwaves are strongly absorbed by water, passing through skin and heating up cells. Microwaves do not cause cancer, but may warm cells when they are absorbed. People are exposed to very low levels of microwave radiation over long periods of time from mobile phones. Many studies have looked for evidence that mobile phones cause harm but have not found evidence of any serious risk. The levels of microwaves that a user is exposed to can be reduced by improving the shielding of the phone, reducing the intensity of the signal or reducing the time of usage.

- Infrared radiation can cause burns.

- Very intense visible light can damage cells in the eye's retina.

- Ultraviolet radiation can cause sunburn in minutes, and prolonged exposure or repeated sunburn can cause skin cancer.

- X-rays and gamma rays can kill or damage cells, or cause burns.

> **Remember!**
> High frequency electromagnetic waves are more harmful than low frequency electromagnetic waves.

Mutations

- Ultraviolet rays, X-rays and gamma rays are ionising. If they are absorbed in cells, they can ionise DNA molecules in the nucleus. The cell may die, or the genetic code may be damaged. The cell may mutate and could form a cancerous tumour.

- The risks increase if the radiation is very intense, or if exposure is over a long period of time or if high-energy waves are absorbed.

Telecommunications

The technological age

- Electromagnetic waves allow signals to travel long distances very rapidly.

- Microwaves communicate with satellites, as they are not absorbed by the atmosphere. Signals are transmitted to the satellite, then transmitted from the satellite to another place on Earth. Microwaves are used for satellite TV broadcasts, sat-nav and mobile phone networks.

- Bluetooth and WiFi networks use low energy radio waves to send signals short distances.

- Visible light and infrared signals travel long distances in optic fibres without being absorbed. Optic fibres form part of the telephone network, also used for Internet and email.

- Interactive TV and TV remote controls use infrared radiation.

Figure 1: How a satellite TV programme reaches you

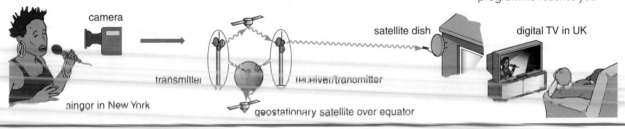

camera · satellite dish · digital TV in UK · transmitter · receiver/transmitter · singer in New York · geostationary satellite over equator

Communications satellites

- Geostationary satellites orbit above the equator at the same rate that the Earth spins so they remain above the same place on Earth. Uses include telecommunication satellites and satellite TV transmission.

> **EXAM TIP**
> You should be able to choose the most suitable type of electromagnetic wave to use in different situations.

Improve your grade

Explain which type of electromagnetic wave is the best choice for satellite communications. **AO2 (3 marks)**

Cable and digital

Why use digital?

- Analogue signals can have any value, but digital signals are pulses that only have two values, on or off.

- Digital signals are higher quality than analogue signals because;
  - there is less interference between different digital signals
  - the digital signal quality is not affected by distance
  - digital signals can be made stronger without losing information.

- Fibre optic cables are thin flexible strands of very pure glass. They transmit information fast with good quality and can carry many signals.

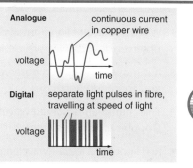

Figure 2: Analogue and digital signals

Total internal reflection

- Total internal reflection is when a light ray repeatedly reflects off a boundary between two different materials. Light rays and infrared signals travel long distances through fibre optic cables because of total internal reflection.

Remember!
Electromagnetic waves can transmit signals using analogue or digital signals.

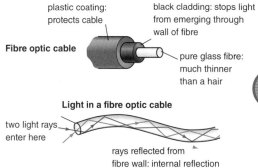

Figure 3: Signals travel through fibre optic cables by total internal reflection

Searching space

Looking further

- Telescopes can be based on Earth or in space.

- Advantages of earth-based telescopes compared to space-based telescopes are;
  - reduced costs to manufacture
  - easier to maintain, repair and update.

- Advantages of space-based telescopes compared with earth-based telescopes are;
  - clearer images as there is no atmospheric interference from moisture, dust, pollution or weather patterns
  - the full spectrum of electromagnetic waves from objects in space can be detected.

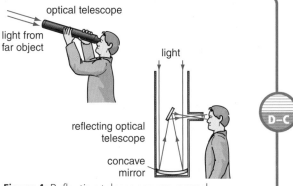

Figure 4: Reflection telescopes use curved mirrors to focus light from stars

Seeing the invisible

- Astronomers can detect a wide range of electromagnetic waves emitted from different objects in space.

Remember!
All types of electromagnetic waves travel at the same speed throughout space.

Radiation	Objects 'seen' in space
gamma ray	neutron stars
X-ray	neutron stars
ultraviolet	hot stars, quasars
visible	stars
infrared	red giants
far infrared	protostars, planets
radio	pulsars

Improve your grade

Explain two advantages of using space based telescopes. **AO2 (4 marks)**

Waves and movement

Doppler and light

- When a source of light, sound or microwaves moves away from an observer, the observed wavelength increases and frequency decreases.

- If it moves towards an observer, the observed wavelength decreases and frequency increases.

- This change in observed wavelength and frequency is called the Doppler effect. It is greater if the source moves faster.

- Red light has a longer wavelength than blue light. Light from galaxies moving away from Earth appears redder (longer wavelength).

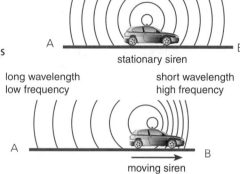

Figure 1: When the siren is stationary both people hear the same sound. When it moves, the person at B hears a higher pitched sound than at A.

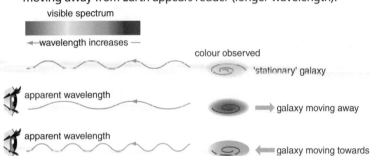

Figure 2: Distant stars and galaxies move fast enough for the colour of their light to change.

Mystery object

- The change in observed wavelength of light provides information on how distant objects move in relation to Earth.

> **Remember!**
> If the wavelength of light or sound increases, its frequency decreases.

Origins of the Universe

CMBR and the Big Bang theory

- Light from many galaxies is shifted towards the red end of the spectrum it appears to have a longer wavelength than light from our own galaxy. This is the red-shift.

- The red-shift is larger when galaxies move away faster. More distant galaxies have a larger red shift. They move away faster.

- The Big Bang theory states that the universe began from a very small initial point where all energy and matter were concentrated. About 14 billion years ago, matter and space expanded violently and rapidly from this point.

- The red-shift is evidence that the Universe is still expanding, supporting the Big Bang theory.

- Cosmic microwave background radiation (CMBR) is electromagnetic radiation filling the universe. It comes from radiation present shortly after the beginning of the universe.

- The Big Bang theory is currently the only theory that can explain the existence of CMBR.

When will the Universe stop expanding?

- Galaxies move apart because of the energy they received at the Big Bang. They are also attracted to each other by gravity.

- If gravity within the universe is great enough, it may stop expanding and reach a fixed size, or collapse back in on itself.

> **Remember!**
> Evidence for the Big Bang includes evidence the Universe is expanding (red shift) and the radiation remaining from the Big Bang (CMBR).

Improve your grade

Explain the evidence we have that supports the Big Bang theory. **AO3 (6 marks)**

P1 Summary

The energy needed to warm 1 kg of a material by 1 degree Celsius is its specific heat capacity.

Solar panels use the Sun's radiation to warm water.

The transfer of energy by heating processes

Kinetic theory can be used to explain different states of matter.

U-values measure how effective materials are as insulators.

Convection transfers heat in liquids and gases; conduction transfers heat most effectively in solids.

The rate of heat transfer depends on surface area, volume, material and type of surface it is in contact with.

Factors that affect the rate at which energy is transferred

All objects absorb and emit infrared radiation. The temperature and colour of the surface affect how quickly this happens.

Conduction, convection, evaporation and condensation transfer energy and involve particles.

Energy can be transferred, stored or dissipated. It cannot be created or destroyed.

The efficiency of a device is the useful energy output / total energy input.

Energy transfers, efficiency and electrical energy

Electrical appliances carry out different energy transfers.

The amount of energy transferred depends on the equipment's power and time it is used for.

Electricity is generated in power stations where heat from a fuel or volcanic rocks changes water to steam. This spins turbines and generators.

Energy from water (hydroelectricity, tides and waves) and wind spins turbines directly. Energy from the sun produces electricity directly.

Generating and distributing electricity

The use of energy resources has an impact on the environment including greenhouse gas emissions, pollution, flooding and waste products.

The National Grid distributes electricity from power stations to consumers. High voltages in cables reduces energy losses.

Waves transfer energy and can be reflected, refracted and diffracted. Their speed is calculated using wavelength × frequency.

Electromagnetic waves are transverse waves that travel at the speed of light in a vacuum and form a continuous spectrum.

The Big Bang theory states that the Universe began as a very small dense and hot point and expanded rapidly. The red shift is evidence the Universe has continued expanding since then.

Waves, communication and the Universe

Radio waves, microwaves, infrared and visible light are used for communication.

Waves are reflected, and this is how an image in a mirror is formed.

Sound waves are longitudinal waves. The loudness and pitch of a note depends on the wave's amplitude and frequency.

See how it moves

Speed and average speed

- Average speed is calculated in m/s using speed = total distance travelled (m)/time taken (s).

Speed from the graph

- Distance time graphs show the distance an object moves in a period of time.

- A line sloping up shows the object moving away from its starting point. If it slopes down, the object is moving back.

- The gradient (slope) at any point gives the speed at that time.

- A steeper gradient shows a faster speed.

- A flat gradient shows the object is stopped.

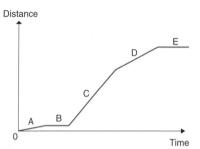

Figure 1: The object is always moving forward, but is stopped at B and E

EXAM TIP

The unit of speed depends on units used for distance and time, for example if distance is measured in km and time is measured in hours, speed is measured in km per hour.

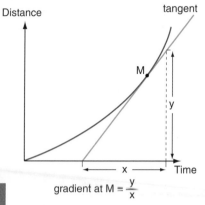

gradient at M $= \dfrac{y}{x}$

Figure 2: Use the gradient of a distance–time graph to calculate speed. If speed changes, the line curves and you must draw a tangent.

Speed is not everything

Velocity and acceleration

- Velocity is an object's speed in a given direction, for example a skydiver falls downwards. A change in velocity (or speed) is called acceleration.

- Acceleration (m/s²) $= \dfrac{\text{change in velocity (m/s)}}{\text{time taken (s)}}$

- The change in velocity is final velocity – original velocity.

- A negative acceleration (deceleration) means the object slows down.

Using velocity time graphs

- Velocity time graphs show how an object's velocity changes with time. The gradient of the graph gives the object's acceleration.

- A steeper gradient means greater acceleration.

- A flat line shows a steady speed.

- The area under a velocity time graph shows the distance travelled. To calculate the area, think of the shape as a combination of triangles and rectangles.

EXAM TIP

It is very easy to confuse velocity time graphs with distance time graphs. Check carefully before writing your answer.

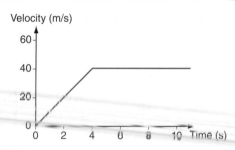

Figure 3: The velocity–time graph shows a falling stone accelerating then travelling at a steady speed

Improve your grade

Sam travels 5 m in 10 s, before stopping for 2 s. He then takes 6 s to return to the start. Draw a distance time graph that shows Sam's journey. **AO2 (3 marks)**

Forcing it

How things move

- The resultant force on an object is the single force that would make an object move in exactly the same way as all the original forces acting together.

- Forces acting in the same direction add, and forces acting in opposite directions subtract.

- If there is no resultant force, forces are balanced. The object does not change speed, so remains stationary.

- If there is a resultant force, stationary objects start moving and moving objects accelerate in the direction of the force.

- If the resultant force is in the opposite direction to motion, the object decelerates.

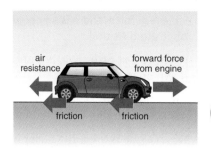

Figure 4: Unbalanced forces on an accelerating car

Using resultant forces

- In a vehicle, fuel consumption increases if drag forces increase at a certain speed. A larger forward force is needed to overcome drag forces. More energy is needed from the engine to provide a larger forward force and keep going at the same speed.

Remember!

The resultant force when a car accelerates is positive (forwards) – drag forces are smaller than the force from the engine. There is no resultant force when the car drives at a steady speed. When a car brakes, the resultant force is negative so the car slows down.

Forces and acceleration

Investigating acceleration

- If there is a resultant force, the object accelerates in the direction of the unbalanced force.

- Acceleration is calculated using: acceleration (m/s²) = $\dfrac{\text{force (N)}}{\text{mass (kg)}}$.

- Acceleration increases as the force applied increases.

- Acceleration decreases if the mass of the object increases.

Mass in space

- The weight of an object is a force that depends on the mass of the object and the gravity acting on it. To find out the mass of an object in space, scientists apply a known force to it and measure its acceleration.

How Science Works

- Streamlined objects have small cross-sectional areas and smooth surfaces. This reduces drag forces so the objects can move faster for the same forward force. Many cars, boats and submarines are streamlined.

Improve your grade

Calculate the total distance travelled by the stone in the first 10 seconds in Figure 3 on page 22. **AO2 (3 marks)**

Balanced forces

Forces on moving objects

- For every force, there is an equal sized force acting in the opposite direction.

- When you sit on a chair, it pushes up on you as hard as you push down on it.

- The weight of a skydiver pulls down on a parachute. This is matched by the upwards tension in the parachute strings.

- When a gun is fired, the bullet feels a forwards force. The bullet exerts an equal sized backwards force on the gun, which recoils.

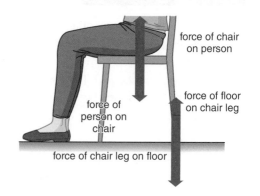

Figure 1: Upwards reaction forces balance the weight acting down

That lifting feeling

- When a lift accelerates upwards, the force from the floor pushing upwards is greater than the weight of the person in the lift. The person feels heavier than normal.

Stop!

Thinking and braking

- Total stopping distance is thinking distance + braking distance.

- Reaction time is the time taken for a driver to react to a hazard.

- Thinking distance = speed × reaction time, and is the distance travelled before the driver reacts and brakes. It increases if the driver:
  - is distracted or tired
  - has taken alcohol or drugs (including some medicines).

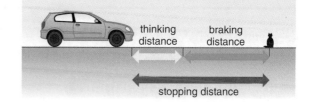

Figure 2: Stopping distance has two parts

- Braking distance is the distance travelled while the brakes are applied and the car is slowing down. It increases if:
  - the road is wet or icy
  - the tyres are worn down
  - the brakes are in bad condition.

Aquaplaning

- Water from the surface of a wet road is channelled out through the treads on a tyre. If the tyre tread is worn, water cannot be removed so the car slides over a layer of water on the road and cannot be controlled. This is called aquaplaning.

How Science Works

- Factors that increase stopping distance also increase the risk of accidents. Drivers who fall asleep at the wheel can be prosecuted for dangerous driving.

Improve your grade

Describe how drag forces compare with forces from the engine for a car that accelerates, then reaches a steady speed then decelerates. **AO2 (4 marks)**

Terminal velocity

Terminal velocity

- Moving objects experience drag forces in the opposite direction to motion. Drag forces increase as an object moves faster.

Figure 3: More air is pushed out of the way at high speed

- The resultant force is the difference between the force causing motion and drag forces.

- A skydiver feels a constant downwards force, weight. Weight is calculated using weight (N) = mass (kg) × gravity (N/kg).

- As the skydiver accelerates, drag forces increase until they match his weight and he reaches a top speed (terminal velocity).

- When the parachute is opened, drag forces increase suddenly. The resultant force is upwards so he decelerates and drag forces decrease.

- Finally he reaches a slower terminal velocity.

Remember!
Skydivers always fall downwards. When the resultant force is upwards, the skydiver decelerates.

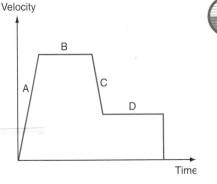

Figure 4: A velocity–time graph for a skydiver

Forces and elasticity

How far does it stretch?

- When an elastic object like a rubber band or spring is stretched:
  - the extension is proportional to the force applied up to a limit
  - above the limit, extension is not proportional to the force.

Remember!
Extension is the difference between the original length and the stretched length.

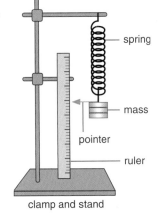

Figure 5: An experiment to test Hooke's law

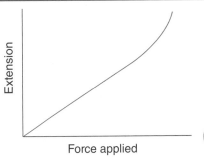

Figure 6: This graph shows Hooke's law

Using elastic potential energy

- Whenever an elastic object is stretched or squashed, work is done to change its shape. The work done is stored as elastic potential energy. This energy can be usefully transferred for example in a catapult or bow, clockwork toy, or small electrical generator.

Improve your grade

Sketch the velocity time graph for a marble that is released into a measuring cylinder of water. The marble reaches its terminal velocity. **AO2 (3 marks)**

Energy to move

Kinetic energy transfers

- A moving object has kinetic energy. This energy has been transferred:
  - from chemical energy in food a person eats
  - from chemical energy in fuel used in an engine.

- The kinetic energy from a moving object is transferred to the surroundings as a result of frictional forces. These include:
  - friction between car tyres and the road surface
  - air resistance felt by aircraft and other moving objects.

- The energy transfer can be useful.
  - Regenerative brakes slow down the car using the engine. The car's kinetic energy is used to charge the car's battery as it slows down.
  - In a car crash, car crumple zones are designed to distort, absorbing kinetic energy.

Storing kinetic energy

- Flywheels are heavy, fast-spinning wheels that store kinetic energy for short periods of time.

Working hard

How much and how far?

- Work is done, and energy is transferred whenever a force moves.

- The amount of work done is calculated using: work done (J) = force (N) × distance moved in the direction of the force (m).

- A person lifts a 60 N parcel by 1.5 m.
  Work done is 60 N × 1.5 m = 90 J.

- A smaller force is needed to drag something up a ramp compared with lifting it directly. The force along the ramp is working over a longer distance, and also needs to work against friction.

> **Remember!**
> The force must move to do work. A person lifting a box does some work but when they stand still just holding a box they are not doing any work.

The clock's stopped!

- When the battery runs out, clocks stop with their moving hands in the "quarter to" position. Most work must be done to lift the hand upwards per tick in this position.

> **EXAM TIP**
> Check diagrams so you use the correct measurements for distance and force when calculating work done.

Improve your grade

a Calculate the work done lifting a piano 0.8 m into a lorry. The piano weighs 1850 N. **AO2 (3 marks)**
b Calculate the force needed to drag the piano into the lorry if a ramp 2 m long is used. **AO2 (2 marks)**

Energy in quantity

Calculating gravitational potential energy

- Gravitational energy is energy an object has because of its position. The object gains more energy when it is lifted higher above the ground.

- Gravitational potential energy (joules) = mass (kg) × gravity (N/kg) × height (m).

D–C

For example, the gravitational potential energy gained by a 3 kg rabbit lifted 0.8 m by its owner is
3 × 10 × 0.8 = 24 J

Remember!
Gravity on Earth is 9.8 N/kg (usually rounded to 10 N/kg).

Remember!
Use the vertical height above the ground when calculating gravitational potential energy.

Investigating kinetic energy

- All moving objects have kinetic energy. Kinetic energy is calculated using:
  - Kinetic energy (J) = ½ × mass (kg) × velocity^2 (m/s)^2
  - for example the kinetic energy of a person (mass 100 kg) running at 3 m/s is ½ × 100 × 3^2 = 450 J
  - The kinetic energy of an object:
    » doubles if its mass doubles
    » increases fourfold if the speed doubles.

- At its highest point, a pendulum has gravitational potential energy and no kinetic energy. As it swings, some gravitational energy changes to kinetic energy and back again.

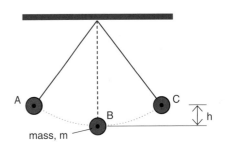

Figure 1: A swinging pendulum changes gravitational energy into kinetic energy and back again

B–A*

Remember!
½ × mass × velocity^2 has a different format to many other equations. Common mistakes include forgetting to multiply by ½, and forgetting that velocity is squared.

How Science Works

- For a moving vehicle, doubling the speed quadruples the kinetic energy and also quadruples the braking distance.

Energy, work and power

Finding the power

- Power measures how quickly work is done, or energy is transferred. It is measured in watts (W) or kilowatts (kW).

- Power is calculated using: power (W) = $\dfrac{\text{energy transferred (J)}}{\text{time (s)}}$ and $\dfrac{\text{force (N) × distance (m)}}{\text{time (s)}}$

D–C

- The power of a motor that does 300 J of work in 10 seconds is 300 J / 10 s = 30 W.

The same work but easier

Remember!
One kilowatt is 1000 W.

B–A*

- People need not pull as hard to drag an object up a ramp compared with lifting it. They only work against a component of gravity, not the direct force.

Improve your grade

Calculate the power of a motor that can lift 600 N in 1 minute. **AO2 (3 marks)**

Momentum

Collisions and explosions

D–C

- All moving objects have momentum.
- Momentum (kgm/s) = mass (kg) × velocity (m/s)
- Momentum has a direction because velocity has magnitude and direction.
- Momentum is conserved. If no external forces act, momentum before and after a collision or explosion is the same.
- In a collision, find the total momentum of moving objects before the collision. The total momentum before and after the collision is the same.
- If a stationary object explodes into two parts, the momentum before and after is zero. Each part after the explosion has the same momentum in opposite directions.

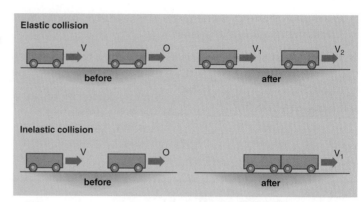

Figure 1: Elastic and inelastic collisions

Remember!
You can choose which velocity direction is positive, but always use a diagram to help you see what is happening in a question.

Jet engines

B–A*

- A jet plane moves because of fast moving compressed air rushing out of the back of a jet engine. The plane gains momentum, moving in the opposite direction to the air.

Figure 2: Air rushing out of the balloon makes it move

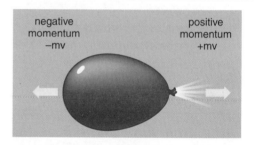

Static electricity

Investigating charges

D–C

- Atoms contain electrons with a negative charge. When different materials are rubbed together, electrons move from one material to the other. Some materials tend to gain electrons and others lose electrons.
- Objects with the same charge repel, objects with opposite charges attract.

Remember!
Electrons are the only particles that move between different materials. Losing electrons makes a material positively charged.

Identifying unknown charges

B–A*

- A gold leaf electroscope with a negative charge can be used to investigate the charge on the other objects. If a negatively charged object comes close, the gold leaf rises more. If a positively charged object comes close, the gold leaf rises less.

How Science Works

- Materials like cling film gain or lose electrons so easily that they just need to touch other materials for an electrostatic charge to build up.

Improve your grade

Two trolleys each with a mass of 1 kg roll towards each other. One trolley travels at 2 m/s to the left and one travels at 3 m/s to the right. They collide and stick together. What is the velocity and direction of the trolleys after the collision? **AO2 (4 marks)**

Moving charges

Electrostatic induction

- A negatively charged object can stick to an uncharged object because of electrostatic induction.

- The negative charge repels electrons from the surface of the uncharged object leaving it with a positive charge.

- The effect is greater on dry days because moisture in damp air can conduct some charge away from the objects.

Figure 3: The balloon sticks because of electrostatic induction

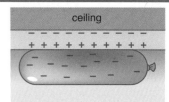

Investigating kinetic energy

- The dangers of electrostatic discharge can be reduced.

- A lightning conductor attracts lightning strikes, which travel down the metal rod to flow safely away.

Circuit diagrams

Series and parallel

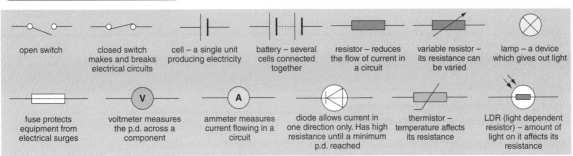

Figure 4: Circuit components

| open switch | closed switch makes and breaks electrical circuits | cell – a single unit producing electricity | battery – several cells connected together | resistor – reduces the flow of current in a circuit | variable resistor – its resistance can be varied | lamp – a device which gives out light |

| fuse protects equipment from electrical surges | voltmeter measures the p.d. across a component | ammeter measures current flowing in a circuit | diode allows current in one direction only. Has high resistance until a minimum p.d. reached | thermistor – temperature affects its resistance | LDR (light dependent resistor) – amount of light on it affects its resistance |

- A series circuit is a loop of conductors. The current is the same throughout the circuit.

- Current is measured in amps using an ammeter connected in series.

- A parallel circuit has components connected in more than one loop.

- The current divides at the junction, and the total current flowing into the junction is the same as the total current flowing out of it.

Series circuit

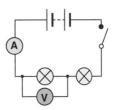

Parallel circuit

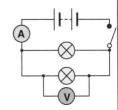

Figure 5: Comparing a series and parallel circuit

Supplying or using energy

- Cells supply energy to the circuit from stored chemical energy in the battery.

- Other components transfer this energy to the surroundings.

- Potential difference is measured in volts using a voltmeter connected in parallel. It measures the energy transferred per unit of charge.

- In series circuits, the potential difference is shared between components.

- In parallel circuits, the potential difference across each loop is the same as the potential difference from the cell.

Improve your grade

Max rubbed a plastic ruler and a metal ruler with a piece of cloth.
a Which ruler became charged? AO1 (1 mark)
b Max held the charged ruler near a charged balloon. The balloon moved away from the ruler.
   Explain why AO2 (2 marks)

Ohm's law

Factors affecting resistance

- Electric current is a flow of moving charges. The charges flow more easily through materials such as copper, because copper has low resistance. If the resistance of a component or wire is high, a small current flows when a given potential difference is applied.

- Resistance is calculated in ohms (Ω) using: resistance (Ω) = voltage (V) ÷ current (A).

- A wire's resistance at a certain temperature is greater if the wire is longer or thinner. This is because electrons are more likely to collide with the nuclei in the material.

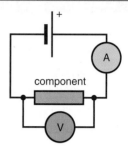

Figure 1: A circuit used to find resistance of a component

Current and potential difference for a resistor

- For some materials, voltage is proportional to current, and resistance is constant. This is called Ohm's law, and applies to ohmic materials.

- A voltage–current graph can be used to find the current in wire at a certain voltage. These values are used to calculate resistance.

EXAM TIP
You may be asked to draw the circuit used to measure resistance.

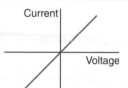

Figure 2: Current–potential difference graph for a resistor at a constant temperature

Non-ohmic devices

Direction, light and temperature

- A filament bulb does not obey Ohm's law. As the current in a filament bulb increases, its resistance increases.

- A diode only allows current to flow in one direction (forward). The arrow on the symbol shows the forward direction. Resistance is very high in the reverse direction.

- A light dependent resistor (LDR) has a smaller resistance when light shines on it.
  - LDRs are used in automatic light systems.

Figure 3: Current–potential difference graph for a filament bulb

- A thermistor has a smaller resistance when it is heated.
  - Thermistors are used in fire alarms and electronic thermometers.

Figure 4: Current–potential difference graph for a diode

How do they work?

- Diodes, LDRs, thermistors and light emitting diodes are made from semiconductor materials. Electrons gain enough energy when the material is heated or light shines on to be able to flow through it.

Remember!
Potential difference is also called voltage, and is measured in volts.

Improve your grade

When the potential difference across a bulb is 6 V, the current through it is 0.1A. What is the resistance of the bulb?
AO2 (3 marks)

Components in series

Potential difference and resistance in series

- In a series circuit the following applies:
  - the current is the same in all places
  - the total resistance of several components is each component's resistance added together
  - the supplied potential difference is shared across components in the same proportion as their resistance. A component with a high resistance has a larger potential difference than a component with a small resistance.

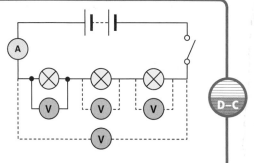

Figure 5: Potential difference in a series circuit

Cells in series

- When you add several cells together, their potential difference adds together.

- If one cell is connected the opposite way to the others, its potential difference is negative.

Remember!
Use voltage = current × resistance to find the voltage across a single component in a circuit.

Components in parallel

Potential difference and resistance in parallel

- The current in a parallel circuit has more than one path. It divides at junctions and joins up again.

- The current in each loop depends on the total resistance in that loop. A smaller current flows through loops with a high resistance compared to loops with a small resistance.

- The total current leaving and returning to the battery is the sum of currents in each loop.

- All bulbs stay as bright as if one bulb was in the circuit.

- The potential difference across each component in a parallel circuit is the same.

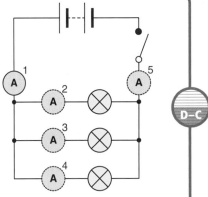

Figure 6: Current in a parallel circuit

Series and parallel

In a loop in a parallel circuit with more than one component:

- the potential difference is shared across the components
- the current is the same through each component.

EXAM TIP
You must be able to use the rules to calculate current and voltage in both series and parallel circuits.

Improve your grade

A series circuit is set up, using four 1.5 V cells in series.
a   What is the total voltage supplied to the circuit?
The circuit includes a motor (resistance 10 ohms) and a bulb (resistance 5 ohms).
b   What is the total resistance in the circuit?
c   Calculate the current in the circuit.
d   What is the potential difference across each component? **AO2 (5 marks)**

Household electricity

Using a cathode ray oscilloscope

- Cells and batteries produce direct current (d.c.), which flows in one direction.

- Alternating current (a.c.) repeatedly changes direction. Mains electricity changes direction 50 times per second (its frequency is 50 hertz) and is about 230 V.

- An oscilloscope can display a.c. or d.c. voltage. Dials set the number of volts per vertical square, and the time per horizontal square.

- The amplitude (maximum height) of the trace shows the maximum potential difference supplied (in volts).

- The number of squares between the peaks shows time per cycle (in seconds).

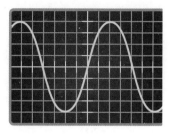

Figure 1: An oscilloscope trace of a.c. current

Calculating period and frequency

- The frequency of the supplied voltage is calculated using:
  frequency (Hertz) = 1/time per cycle (seconds).

  For example, if the frequency is 50 Hz, the time per cycle is $\frac{1}{50}$ = 0.02 s.

Remember!
Amplitude is measured from the centre of a trace to the highest (or lowest) point.

Plugs and cables

Cables and fuses

- In a three-pin plug, wires are covered in colour-coded plastic:
  - the neutral wire is blue
  - the live wire is brown
  - the earth wire is green and yellow.

- Appliances with a plastic outer case and no touchable metal parts are double insulated. They use two-core cable (live and neutral wires).

- Other appliances use three-core wire (earth, live and neutral wires). If there is a fault and the equipment becomes live, the current flows through the earth wire and blows the fuse.

- The fuse is connected in series with the live wire. It protects the appliance and flex from overheating. If the current is too large, the fuse melts and breaks the circuit.

Figure 2: Inside a three-pin plug

EXAM TIP

Make sure you can describe dangerous habits when using electricity and spot mistakes when wiring plugs.

Figure 3: Three-core cable

Shocking

- Electric shocks can kill. Wet skin has lower resistance than dry skin so electrocution is more likely.

- Mistakes when wiring a plug include:
  - connecting wires to the wrong pins
  - not gripping the cable tightly under the cable grip
  - not attaching wires firmly to the pins
  - having too much bare wire exposed.

EXAM TIP

You should be able to explain why certain materials are used for different parts of a plug and cable.

Improve your grade

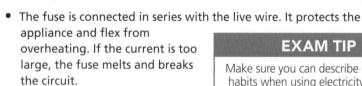

Explain why a double insulated appliance does not need an earth wire. **AO1 (2 marks)**

Electrical safety

Residual current circuit breakers (RCCBs)

- RCCBs protect users from electrocution. The current in the live and neutral wires should always be the same. Residual current circuit breakers (RCCBs) break the circuit in less than 0.05 seconds if there is a difference in the current in these wires. The RCCB can be reset once the fault is repaired.

- Many homes now have RCCBs in them.

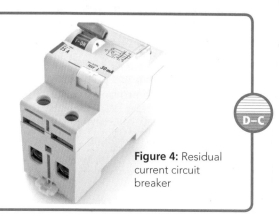

Figure 4: Residual current circuit breaker

Choosing the right fuse

- Fuses are rated in amps, for example 3 A, 13 A.

- If the fuse rating is too low, the fuse melts even if the appliance works correctly. If it is too high, it won't melt if there is a fault.

- Use current (A) = $\dfrac{\text{power (W)}}{\text{voltage (V)}}$ to calculate which fuse to use.

How Science Works

- Mains voltage is supplied at 230 V so a 3 A fuse only works if the power of the equipment is 690 W or less. Otherwise, a 13 A fuse is needed.

Current, charge and power

Current charge and work done

- An electric current transfers energy.

- The energy transferred is calculated using: energy transferred (J) = power (W) × time (s).

- The electric current is a flow of charge (electrons). Charge is measured in coulombs, and has the symbol Q.

- Current is the rate that charge flows. It is calculated using: current (A) = charge (C)/time (s).

Calculating the energy transferred

- A battery supplies charge with energy. One volt supplies each coulomb of charge with one joule of energy.

- Calculate the energy supplied using: energy (J) = potential difference (V) × charge (C)

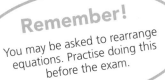

Remember!
You may be asked to rearrange equations. Practise doing this before the exam.

EXAM TIP

When choosing the equation to use, check what you are asked for and what you are given in the question.

Improve your grade

Fuses rated at 3 A, 5 A and 13 A are available. Which fuse should be used in each case?
a a lawn mower (power 2000 W)
b a lamp (power 60 W)
c a toaster (power 800 W) **AO2 (3 marks)**

Structure of atoms

Electrons, protons and neutrons

- Atoms are made from particles called electrons, protons and neutrons, which have different masses and charges. The atomic number is the number of protons in the atom. The mass number is the number of neutrons and protons.

Particle	Charge	Relative mass
neutron	0	1
proton	1	1
electron	−1	negligible

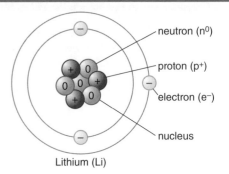

Lithium (Li)

Figure 1: An atom of lithium

- Atoms have no overall charge as they have equal numbers of electrons and protons. An atom that loses electrons becomes a positive ion. An atom that gains electrons becomes a negative ion.

- All atoms of the same element have the same number of protons, but can have different numbers of neutrons. Atoms of the same element with different numbers of neutrons are called isotopes.

- Unstable atoms have different numbers of protons and neutrons. Some will decay emitting ionising radiation from the nucleus.

Subatomic particles

- The model of the atoms is evolving. Scientists are looking for experimental evidence of unknown particles, and the forces that hold them together.

Remember!
There are a lot of new terms in this topic. Learn their definitions carefully.

Radioactivity

Explaining the properties

- Radioactive atoms emit ionising radiation from the nucleus. Atoms decay at random times. An atom that absorbs ionising radiation may change into an ion. Types of ionising radiation include:
  - Alpha particles which consist of 2 neutrons and 2 protons (a helium nucleus).
  - Beta particles which are electrons emitted when a neutron changes to a proton.
  - Gamma radiation which is high energy electromagnetic radiation.

Nuclear equations

- The mass number is the number of protons and neutrons in an atom. The proton number is the number of protons in an atom.

- Nuclear equations show how isotopes decay by emitting alpha and beta radiation. The mass number and the proton number on each side of the equation must balance.

- This equation shows alpha decay

$$^{224}_{88}\text{Ra} \rightarrow {}^{220}_{86}\text{Rn} + {}^{4}_{2}\text{He}$$

- This equation shows beta decay

$$^{212}_{83}\text{Bi} \rightarrow {}^{212}_{84}\text{Po} + {}^{0}_{-1}\text{e}$$

EXAM TIP

Make sure you know the properties of nuclear radiation and can use them to explain how the radiation affects nearby atoms and molecules.

Improve your grade

Why are the risks from handling an alpha source safely different from the risks from handling a gamma source?
AO3 (3 marks)

More about nuclear radiation

Explaining the scattering experiment

- The plum pudding model was an early model of the atom. It was not a successful model so a scientist called Rutherford carried out an experiment to explain the structure of an atom.

- He fired alpha particles at gold foil and observed the scattering of the particles.

- Most alpha particles passed straight through as atoms were mainly empty space.

- Positively charged alpha particles were deflected by positive charge concentrated in the nucleus.

- A few massive alpha particles were reflected by the very dense nucleus.

- The pattern helped to explain the structure of atoms. Rutherford's nuclear model replaced the plum pudding model.

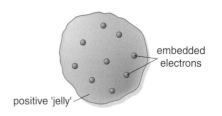

Figure 2: The plum pudding model.

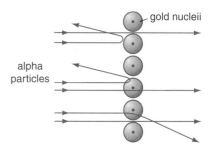

Figure 3: Explaining the alpha particle tracks.

More deflections

- Alpha and beta particles have opposite charges so they are deflected in opposite directions by electric and magnetic fields.

- Alpha particles are more massive than beta particles so they are deflected less.

Background radiation

Natural and lifestyle

- Background radiation is radiation that is all around us and comes mainly from natural sources.

- Natural sources include rocks, soil and food.

- Sources are affected by lifestyle choices, for example, on a long flight you increase your exposure to cosmic radiation.

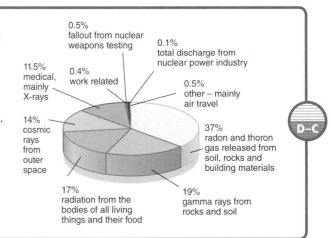

Figure 4: Sources of background radiation.

Dangerous levels – or not?

- It is hard to assess the risks from background radiation as many different factors affect your lifetime cancer risk.

Improve your grade

Explain how the Rutherford scattering experiment provided evidence for the nuclear model of the atom. **AO2 (3 marks)**

Half-life

Measuring half-life

- We cannot predict if a specific radioactive atom will decay, but we can predict how many atoms will decay in a sample in a given time.

- Half-life is the time taken for the original radioactivity or count rate of a sample to halve.

- After a time equal to two half lives, the activity falls to a quarter of its original value.

- Each radioactive material has its own half-life, which varies from millions of years to milliseconds. The half-life does not change for a given material.

Graphs to identity radioactive isotopes

- The graph shows how the activity of a sample varies with time.

To find the half-life;

- work out the value of half the original count rate

- use the graph to read the time taken for the activity to reach this level.

Remember!
Whenever you start counting and whatever the original count rate, the count rate or activity falls to a half of the original value after one half-life.

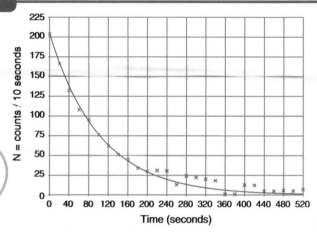

Figure 1: Radioactive decay graph for strontium-93

Using nuclear radiation

Using radioisotopes

- Radioisotopes are radioactive isotopes. They have different uses.

- Gamma rays kill living cells and penetrate materials so are used for sterilisation and cancer treatments.

- Beta radiation penetrates thin sheets of metal and cardboard so is used to monitor the thickness of materials.

- Medical tracers are radioactive materials injected, inhaled, or eaten by a patient. The position of the radioactivity is monitored to check blood flow, or identify blockages or tumours.

- Alpha radiation is used in smoke detectors. The particles ionise air so a current can flow. If smoke absorbs the radiation, the current cannot flow switching the alarm on.

Remember!
Radioactive isotopes with a short half-life are used with living organisms to reduce harm. In monitoring equipment, isotopes with a long half-life are used to maintain consistent readings.

Plant growth

- Radioactive water emits beta radiation so it can be used to monitor plant growth.

Improve your grade

The half-life of one material is 20 days, and its original activity is 100 000 counts per minute. Another sample has a half-life of 15 days, and an original count rate of 200 000 counts per minute. What is the count rate of each sample after 60 days? **AO2 (3 marks)**

Nuclear fission

Critical mass and chain reaction

Nuclear fission is when a large nucleus splits into two or more parts. In nuclear power stations:

- uranium-235 and plutionium-239 are used as fuels,

- atoms of uranium or plutonium absorb an extra neutron and become unstable,

- energy is released when the nucleus splits into two parts plus two or three neutrons.

- more uranium or plutonium atoms may absorb the extra neutrons. This can start a chain reaction involving more atoms at each stage.

- Control rods in nuclear power stations control the size of the chain reaction by absorbing surplus neutrons.

- The critical mass of a nuclear fuel is the minimum amount needed to keep the chain reaction going.

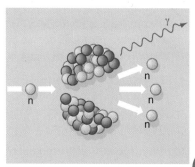
Figure 2: Splitting the atom

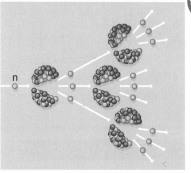
Figure 3: A chain reaction

D–C

Why is so much energy released?

- Strong forces in the nucleus mean that large amounts of energy are released when the atom is split. Some of the mass of the nucleus is changed into energy during nuclear fission.

EXAM TIP

You should be able to describe the stages taking place during nuclear fission.

B–A*

Nuclear fusion

Using nuclear fusion

- Nuclear fusion is when two small nuclei join together forming a nucleus of a different element and releasing energy. This is how energy is released in stars.

- Very high temperatures (above 15 million degrees Celsius) and very high pressures are needed for nuclear fusion because positively charged protons in nuclei repel each other.

- Nuclear fusion in stars produces elements lighter than iron.

- Elements heavier than iron are produced during supernova (when massive stars explode releasing huge amounts of energy).

D–C

The proton-proton chain

- Nuclear fusion of hydrogen takes place in three stages in stars.

EXAM TIP

Be very careful not to confuse nuclear fission and nuclear fusion, and learn the correct spellings of each.

B–A*

Improve your grade

Explain why the presence of iron in the Sun is evidence that the Sun formed from the remains of older stars.
AO2 (3 marks)

Life cycle of stars

Why does the main sequence last so long?

- All stars have a life cycle. When gravity pulls dust and gas together, a protostar forms. Smaller masses (planets) may also form. The life cycle of a star is determined by its size.

- The star's main sequence lasts billions of years. Nuclear fusion changes hydrogen to helium, and heavier nuclei. The star is stable because forces are balanced within the star (inward gravitational forces match outward forces due to the heat).

- As hydrogen fuel is used up, nuclear fusion reactions change. The star expands becoming a red giant.

- When nuclear fusion reactions change again, the inner core collapses. The star becomes a white dwarf. Nuclear reactions form elements up to iron.

- When its nuclear fuel runs out, the star cools and becomes a black dwarf.

D–C

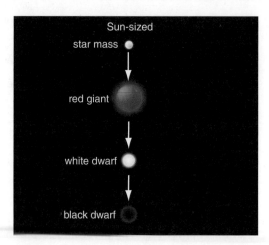

Figure 1: The life cycle of a star like the Sun

A matter of size

- Stars much bigger than the Sun become a red super giant.

- When its fuel is used up, the core collapses quickly and the star explodes as a supernova. Elements heavier than iron form in a supernova.

- After the supernova, stars up to 3 times heavier than the Sun become neutron stars. Heavier stars create a black hole.

B–A*

Remember!
You should be able to explain how changes in the forces acting on a star make it move between different stages.

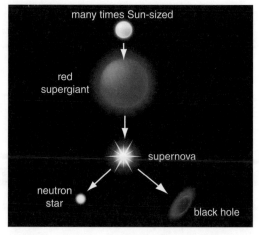

Figure 2: The life cycle of a star much bigger than the Sun

Improve your grade

Explain how the stages in a star's life cycle are controlled by its mass. **AO2 (3 marks)**

P2 Summary

A resultant force shows the combined effect of several forces. It makes an object accelerate (change speed or direction).

The stopping distance of a vehicle is thinking distance plus braking distance.

Moving objects reach a terminal velocity (top speed) when drag forces matches driving forces.

Forces and energy

The speed of an object can be found using a distance time graph. The acceleration and distance traveled can be found using a velocity time graph.

Momentum is mass × velocity. It is conserved when objects collide or explode apart if no external forces act.

A force transfers energy when it does some work on an object.

Objects gain gravitational potential energy if they are lifted up. Moving objects have kinetic energy. Stretched or squashed elastic objects store elastic energy.

Insulating materials can be charged by rubbing. They gain or lose electrons.

Static electricity

Objects carrying the same charge repel. Objects carrying the opposite charge attract.

Electric current is a flow of charge.

Electrical circuits

Resistance measures how easily a current flows through a component.
Resistance = potential difference ÷ current

In a series circuit: the current is the same throughout, the potential difference is shared between components and the resistance of components adds.

Circuit diagrams use standard symbols.

In a parallel circuit, the current is shared between branches and the potential difference is the same for each branch.

Batteries supply direct current (d.c.); mains supply is alternating current (a.c.) at about 230V 50Hz.

Using mains electricity safely and the power of electrical appliances

Earthing, fuses and RCCBs protect the user and appliance.

Electric cable can be two-core or three-core. A three pin plug must be correctly wired.

Power is the potential difference supplied × current, or energy transferred ÷ time.

Energy transferred is the potential difference supplied × charge.

An atomic nucleus contains protons and neutrons. Electrons orbit the nucleus. Radioactive materials emit ionising radiation from their nucleus.

Atomic structure and radioactivity

Background radiation comes from natural sources (for example rocks,) and man-made sources (for example medical uses).

Half-life is the time taken for the count rate of a radioactive sample to halve.

Alpha, beta and gamma radiation have different properties, uses and dangers.

Ions are atoms that have lost or gained electrons. Isotopes have the same number of protons, but different numbers of neutrons.

Nuclear fission is the splitting of an atomic nucleus.

Nuclear fission and nuclear fusion

Nuclear fusion is when two atomic nuclei join to form a larger one.

Fission of uranium-235 or plutonium-239 during a chain reaction releases energy in nuclear reactors.

All stars have a life cycle. Nuclear fusion in stars releases energy and produces all naturally occurring elements.

Using X-rays

Using X-rays

- X-rays are part of the electromagnetic spectrum. They cause ionisation because they are high-frequency, high-energy waves.

- X-rays are absorbed by dense materials like bone or metal, but pass through less dense materials like paper and healthy tissues.

- X-rays are used:
  - in industry, to examine materials without damaging them
  - in airports, to examine luggage for security reasons
  - in medicine, to diagnose medical conditions.

- Medical X-rays have enough energy to pass through all tissues except teeth and bones. If soft tissues absorb barium or iodine, they will absorb X-rays.

- Computerised tomography (CT or CAT scans) combine many X-rays taken from slightly different positions to produce a 3-D image of the body.

- Cancer cells are killed using focused beams of X-rays.

- X-rays are detected using photographic film, or electronic detectors that turn the X-rays into an electronic image that can be seen on a screen.

Remember!
X-rays produce shadow pictures. Only X-rays that are not absorbed by the material can pass through to the detectors.

X-rays and safety

- X-rays ionise cells and increase the risk of cells becoming cancerous. The risk increases with exposure, so a single X-ray is unlikely to be harmful.

- People working with X-rays are protected using lead screens. Film badges monitor their exposure to X-rays.

Ultrasound

Using ultrasound

- Ultrasound is sound waves with a frequency above 20,000 Hz (too high for human hearing). Ultrasound travels at different speeds in different materials and tissues.

- Pulses of ultrasound are used to measure distance inside materials because they partially reflect at boundaries between different materials.

- The speed of ultrasound in different materials is known.

- The time for the pulses to reflect off the boundary and return to the detector is measured.

- The distance to the boundary and back is calculated using speed × time.

- The distance to the boundary is half this amount.

- Medical uses of ultrasound include:
  - scanning fetuses to check their development
  - detecting and treating medical conditions such as kidney stones.

Remember!
You can calculate the time using an oscilloscope trace. Multiply the number of squares between pulses by the time that each square represents.

Skin contact

- A gel is smeared on skin before an ultrasound examination. This stops pulses reflecting off the boundary between skin and air.

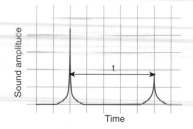

Figure 1: An oscilloscope trace for an ultrasound pulse and its echo

Improve your grade

Explain why ultrasound scans are safer to use than X-rays. **AO2 (2 marks)**

Refraction

The power of lenses

- Refraction is when light changes direction as it passes from one material into another.

- You calculate the refractive index using:

  refractive index = sin i / sin r

  - i is the angle of incidence (between the normal and incident ray)
  - r is the angle of refraction (between the normal and refracted ray).

- The refractive index does not have a unit.

- Lenses use refraction to form an image.

- Parallel rays of light come together (focus) at a point called the principal focus.

- The distance between the lens and the principal focus is the focal length, f. Focal length is measured in metres, m.

- The power of the lens is 1/f. Power is measured in dioptres, D.

- Convex (converging) lenses bring the light together.

- Concave (diverging) lenses spread light out.

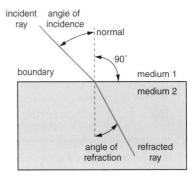

Figure 2: Refraction in a glass block

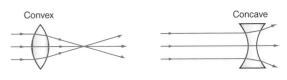

Figure 3: Convex and concave lenses can be represented by symbols

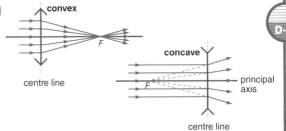

Figure 4: Ray diagrams for convex and concave lenses

Remember!
Refraction happens at the boundary between the materials. Light bends away from the normal when it passes into less dense materials (like air).

Making thinner lenses

- If a lens needs to have a shorter focal length:
  - it can be made more curved
  - it can keep the same shape but be made of a material with a larger refractive index.

Remember!
Images from concave lenses are always upright, virtual and diminished (smaller).

Improve your grade

Explain how a lens can be used as a magnifying glass. **AO2 (3 marks)**

Lenses

Magnification

- Ray diagrams show the position and size of an image. You must draw these accurately to scale.
  1. Draw the principal axis through the centre of the lens.
  2. Mark the focal point, F, and the position of the object, drawn to scale.
  3. Draw one ray from the top of the object through the centre of the lens without changing direction.
  4. Draw a second ray from the top of the object parallel to the principal axis. When it reaches the lens, the ray passes through the focal point.
  5. Extend both rays till they meet at the top of the image.

- The magnification of a lens is calculated using image height / object height. It has no unit.

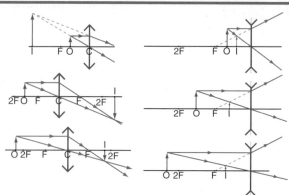

Figure 1: Ray diagrams for a convex lens

Figure 2: Ray diagrams for a concave lens

Images

- Images can be:
  - magnified (larger than the object) or diminished (smaller than the object)
  - upright or inverted (upside down)
  - real (formed the opposite side of the lens as the object) or virtual (formed the same side of the lens as the object).

Seeing clearly

Comparing an eye with a camera

- Parts of the eye include:
  - the retina, which has light-sensitive cells that send signals along the optic nerve to the brain
  - the lens, which changes shape to focus light from objects at different distances onto the retina
  - the cornea, which is a clear covering that protects the eye
  - the pupil, which allows light into the eye
  - the iris, which changes size to change the pupil size
  - the ciliary muscles, which pull on the lens to change its shape
  - the suspensory ligaments, which link the lens to the ciliary muscles.

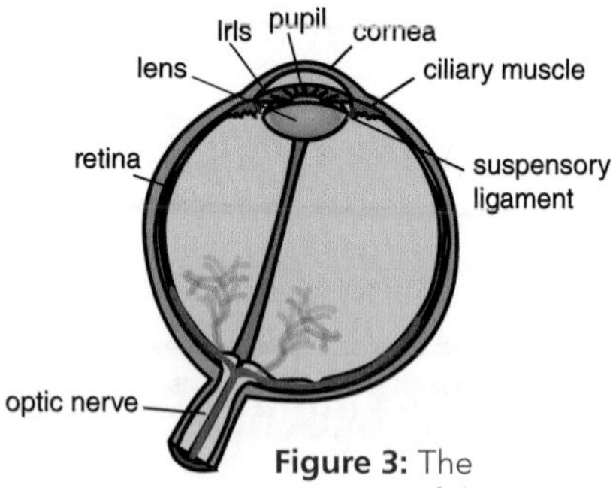

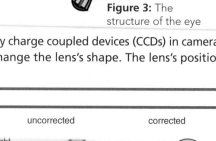

Figure 3: The structure of the eye

- The eye and a camera have similarities and differences.
  - Light is detected by the retina in the eye, and by photographic film or by charge coupled devices (CCDs) in cameras.
  - To focus on objects at different distances, ciliary muscles in the eye change the lens's shape. The lens's position is adjusted in the camera.

Correcting vision

- People who are short sighted do not see distant objects clearly:
  - the lens focuses light in front of the retina
  - to correct this, a concave lens is needed to spread light from distant objects out
  - the image then focuses further back on the retina.

- People who are long sighted do not see nearby objects clearly:
  - the lens focuses light behind the retina
  - to correct this, a convex lens is used to make light from nearby objects converge
  - the image then focuses further forward on the retina.

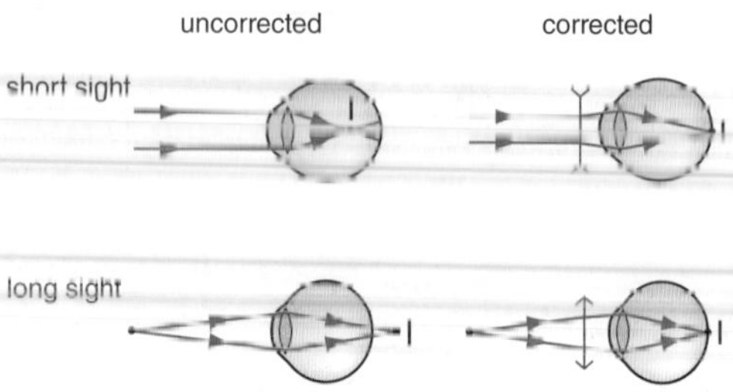

Figure 4: Correcting long- and short-sightedness

Improve your grade

Describe the changes taking place in the eye when a person focuses on a nearby object then on a distant object.
AO2 (4 marks)

More uses of light

Total internal reflection

- Optic fibres are made from thin flexible glass. Visible light travels through the fibre by repeatedly reflecting off its inside surfaces. This effect is called total internal reflection.
  - Total internal reflection occurs when the angle of incidence is higher than the critical angle.

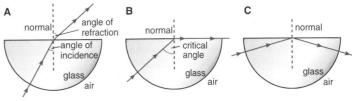

Figure 5: Total internal reflection in a semicircular glass block

- An endoscope uses optic fibres and a camera to examine inside patients without an operation. Visible light sent through the optic fibre reflects off internal organs to the camera.

- Lasers are very intense sources of single-coloured light. They are an energy source that can be used for:
  - cutting (for example, during surgery)
  - cauterising (sealing cut blood vessels)
  - burning (for example, reshaping the cornea to correct short sight; repairing damaged retinas by welding tissues back together).

How Science Works

- Cats eyes in the road use total internal reflection to reflect light from headlights back to the driver's eyes.

D–C

Critical angle

- The critical angle of a material depends on its refractive index. It is found using this equation:

$$\text{refractive index} = \frac{1}{\sin c}$$

B–A*

Centre of mass

Finding the centre of mass

- The centre of mass is the point where the mass of an object is thought to be concentrated. It is along the axis of symmetry for a symmetrical object.

- If you suspend an object, its centre of mass lies directly below the point of suspension.

- Objects topple if their centre of mass does not lie over the base.

- Stable objects may have a wide base, be short, or have their mass concentrated at the base.

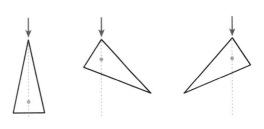

Figure 6: The centre of mass always lies under the point of suspension

- To find the centre of mass for an object that is not a regular shape:
  - suspend it from three points around its edge
  - mark a vertical line directly under each point of suspension
  - the centre of mass is where these points cross.

D–C

Pendulums

- A pendulum swings freely and regularly from its point of suspension.
- The time period for one complete swing, T, is found using: $T = \frac{1}{f}$
  - T is measured in seconds, and the frequency, f, is measured in hertz (Hz).
- A longer pendulum has a larger time period.

Remember!
The time period does not depend on the mass of the object or amplitude of the swing.

B–A*

Improve your grade

Explain why double decker buses are less stable than single deck buses. AO2 (2 marks)

Moments

Balancing moments

- The turning effect of a force is called its moment.
- The size of the moment is found using: moment = force × perpendicular distance from pivot
- Moments are measured in Newton-metres.
- Moments are balanced and the object does not turn if moments in the clockwise and anticlockwise directions match.
- Levers change the size of force needed to make something happen. Increasing the length of a lever reduces the force needed.

Tendency to topple

- The seesaw in Figure 1 balances if $F1 \times d1 = F2 \times d2$
- For a balanced object, you can use this equation to calculate an unknown force or distance.
- Objects topple if their centre of mass lies outside the base of the object. The moment caused by the object's weight makes the object turn.

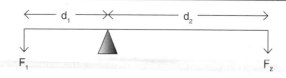

Figure 1: Calculating moments

EXAM TIP

Questions on moments often include diagrams. Use the diagram to find where the pivot is, and the forces and distances involved.

Hydraulics

Calculating the force

- Liquids are virtually incompressible. A pressure in the liquid is transmitted equally in all directions. Hydraulic systems use liquids in pipes to transmit a force.

force in

force out

Figure 2: A simple hydraulic system

- In a hydraulic system:
  - a force (effort) on a piston exerts a pressure in the liquid
  - the pressure is the same at all points in the liquid
  - this force is transferred to a piston at the other end of the hydraulic system
  - changing the area of the pistons changes the size of the outward force (load).
- The pressure in a hydraulic system is found using: $P = \dfrac{F}{A}$
  - where P is pressure in Pascals (Pa); F is force in newtons (N) and A is the area of the piston in m^2.
- A hydraulic system acts as a force multiplier. For the piston shown in Figure 3:

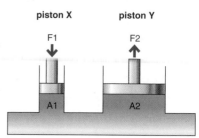

piston X   piston Y

F1   F2

A1   A2

Figure 3: A simple hydraulic system

Pressure at X = Pressure at Y

$$\frac{F_1}{A_1} = \frac{F_2}{A_2}$$

Conservation of energy

- Energy transferred is force × distance moved by the force.
- A hydraulic system can increase a force, but never increases the energy transferred.
- The smaller piston moves a larger distance than the larger piston.

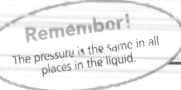

Remember!
The pressure is the same in all places in the liquid.

Improve your grade

Calculate the force, F2, in Figure 1 if F1 is 30 N, d1 is 1 m and d2 is 2 m. **AO2 (2 marks)**

Circular motion

Centripetal force

- When an object changes speed or direction there is a resultant force. When an object moves in a circle:
  - its direction constantly changes, so its velocity constantly changes (velocity is speed with direction)
  - the object constantly accelerates, so it feels a constant force
  - the force is called a centripetal force.

- Examples of centripetal forces are gravity between planets and the Sun; friction between car tyres and the road surface; and tension in a string. The centripetal force increases if:
  - the speed increases
  - the mass increases
  - the radius of the circle decreases.

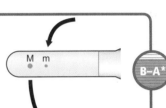

Figure 4: Centripetal force acting on a mass

D–C

Centrifuges

- Centrifuges separate a mixture of liquids of different densities by spinning them.

- Heavier particles move to the furthest point of the centrifuge tube.

Remember!
Centripetal force always acts towards the centre of the circle, stopping an object travelling in a straight line.

Figure 5: Separating particles in a centrifuge

B–A*

Circular motion in action

A simple pendulum

- Gravity provides a centripetal force that keeps planets and satellites in orbit. Gravity is stronger between objects that are closer together so these objects travel faster in their orbit.

- A pendulum is a mass on a string that can swing freely. Examples include a swing, and some fairground rides.

- At the highest point of its swing, the pendulum briefly stops and stores energy as gravitational potential energy.

- At the lowest point of the swing, this energy is transferred to kinetic energy as the pendulum travels fastest.

- The time period of a pendulum, T, is calculated using $T = \dfrac{1}{f}$

  - where T is the periodic time in seconds and f is the frequency in hertz.

- The longer the pendulum, the longer its time period.

Figure 6: Satellites in a polar orbit travel faster than satellites in a geostationary orbit

D–C

Long pendulums

- Changing the mass of a pendulum does not change the time period of a pendulum.

B–A*

EXAM TIP

You may need to use ideas about circular motion to describe how different pendulums behave.

Improve your grade

Explain how the centripetal changes in each case:
a A car slows down as it drives round a bend
b A bucket of water is swung in circles when it is full of water and when it is empty. **AO1 (2 marks)**

Electromagnetic force

Building an electric motor

- When a current flows in a wire:
  - a magnetic field is produced around the wire creating an electromagnet
  - if the wire is near a magnet, it feels a force and the wire moves. This is called the motor effect.

- There is no force (or motor effect) if the current is parallel to the magnetic field.

- The size of the motor effect increases if:
  - the current increases
  - a stronger magnet is used.

- An electric motor uses the motor effect to spin a coil of wire. The motor turns faster if the magnetic field or current increases. The direction of the motor effect changes (and the motor spins the other way) if either:
  - the direction of the current changes
  - the direction of the magnetic field changes.

- Fleming's left-hand rule identifies the direction of a force if the current and magnetic field directions are known.

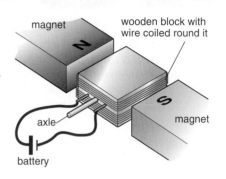

Figure 1: A simple electric motor

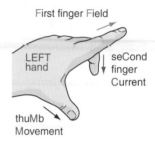

Figure 2: Fleming's left-hand rule

Electromagnetic induction

Coils for magnets

- When a wire in a complete circuit moves across a magnetic field:
  - a potential difference is induced across the wire
  - a current flows in the circuit
  - this is electromagnetic induction.

- Electromagnetic induction takes place whether the magnet or the wire moves.

- The current is larger if any of these changes are made:
  - the magnetic field is stronger
  - the magnet or coil moves faster
  - there are more turns in the wire coil
  - the wire coil has a larger diameter.

- If the magnet is replaced with a coil carrying a current, a current is still induced so long as either coil is moving.

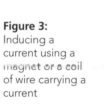

Figure 3: Inducing a current using a magnet or a coil of wire carrying a current

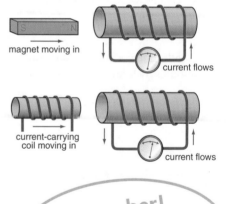

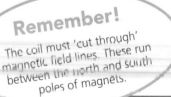

Remember!
The coil must 'cut through' magnetic field lines. These run between the north and south poles of magnets.

Surge protection

- Sensitive equipment like computers must be protected against surges of current. The surges happen because the current changes very quickly when equipment is turned on or off. The induced current can cause a lot of damage.

Improve your grade

Explain how the motor effect is used to make the drum in a washing machine spin. **AO2 (3 marks)**

Transformers

How transformers work

- Transformers are devices that use electromagnetic induction to increase and decrease the potential difference in a circuit. They only work with alternating current (a.c.). They are used for devices such as mobile phone chargers that do not use 230 V mains supply.

- Transformers are made from two insulated coils of wire wrapped around an iron core.
  - An a.c. current flowing in one coil (the primary coil) sets up a changing magnetic field in the soft iron core.
  - The changing magnetic field in the iron core induces an a.c. voltage in the other coil (the secondary coil).
  - This means an a.c. current flows in the secondary coil.

- A step-up transformer increases potential difference. There are more turns on the secondary coil than the primary coil.

- A step-down transformer reduces potential difference. There are more turns on the primary coil than the secondary coil.

- The potential differences across the coils in a transformer are related, this is shown in the turns rule: $\frac{V_p}{V_s} = \frac{n_p}{n_s}$ where:

  V_p is the potential difference across the primary coil
  V_s is the potential difference across the secondary coil
  n_p is the number of turns in the primary coil
  n_s is the number of turns in the secondary coil.

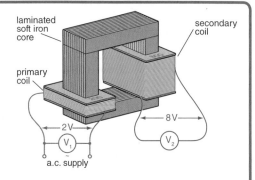

Figure 4: The structure of a transformer

> **Remember!**
> The coils are insulated. No current flows through the iron core.

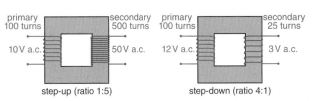

Figure 5: Step-up and step-down transformers

step-up (ratio 1:5)　　step-down (ratio 4:1)

Using transformers

Transformers in use

- In many calculations we can assume that transformers are 100% efficient. The electrical power input is the same as the electrical power output.

- This means that $V_p \times I_p = V_s \times I_s$ where:
  - V_p = potential difference across the primary coil in volts, V
  - I_p = current through the primary coil in amps, A
  - V_s = potential difference across the secondary coil in volts, V
  - I_s = current through the secondary coil in amps, A.

> **Remember!**
> You may be asked to link the ideas about electromagnetic induction to explain how a transformer works.

- Transformers are not 100% efficient, and become warm when they are turned on. Switch-mode transformers warm up less than traditional transformers, because they use very little power if they are switched on but no load is applied to them (for example, plugged in but the equipment is not being used).

- Traditional transformers operate at 50 Hz, which matches mains supply frequency. Switch-mode transformers are small, light transformers. They operate at very high frequencies (between 50 kHz and 200 kHz).

How switch-mode transformers work

- The high frequency current in a switch-mode transformer induces a larger voltage than a current at a lower frequency. This means switch-mode transformers can use a smaller iron core than traditional transformers.

Improve your grade

A transformer has 3000 turns on its primary coil and 6000 turns on its secondary coil.
a Is it a step up or a step down transformer? **AO1 (1 mark)**
b What is the voltage across the secondary coil if the voltage across the primary coil is 12 V? **AO2 (2 marks)**

X-rays and CT scans diagnose and treat medical conditions. Precautions are necessary as they can cause ionisation.

X-rays and ultrasound

Ultrasound waves are very high frequency sound waves. They are partly reflected at boundaries. The distance s to a boundary is $v \times t$.

Refraction is when light changes direction as it passes into a different medium.

Lenses and the eye

Refractive index is calculated using sin i/sin r or 1/sin c where c is the critical angle.

Lenses are converging (convex) or diverging (concave). Images are real or virtual, upright or inverted, magnified or diminished.

Ray diagrams show how images are formed. Magnification is calculated using image height/object height.

Convex lenses correct long sight, and concave lenses correct short sight. A lens' power is calculated using 1/f. It is affected by the shape and material of the lens.

Structures in the eye include the retina, lens, cornea, pupil/iris, ciliary muscle and suspensory ligaments. The camera has similar structures.

Visible light travels along optical fibres because of total internal reflection.

Other applications of light

Lasers are used for cutting, cauterising and burning.

We can say an object's mass is concentrated at its centre of mass. This hangs beneath a point of suspension, or lies on an axis of symmetry.

Centre of mass and moments

The moment of a force is its turning effect. It is calculated using force × distance from pivot. Clockwise and anticlockwise moments balance if the object does not turn. This is used to calculate a force or its distance from a pivot.

The time period, T, of a pendulum is 1/frequency, f.

Objects topple if there is a resultant moment.

Levers act as force multipliers.

Liquids are incompressible and transmit forces in all directions.

Hydraulics

The pressure in a hydraulic system is P = force/area. A hydraulic system can multiply forces.

Circular motion

Objects moving in a circle accelerate towards its centre. The centripetal force is larger if the mass or its speed increases, or the radius of the circle decreases.

A current flowing in a wire creates a magnetic field. This causes a force (the motor effect) if the wire is in a magnetic field.

The motor effect

The motor effect increases in a stronger magnetic field or if the current is larger. It reverses if the direction of field or current reverses.

An alternating p.d. in the primary coil induces an alternating p.d. in the secondary coil. A step-up transformer increases the p.d.; a step-down transformer reduces the p.d. The ratio of p.d.s in the coils is the same as the ratio of the number of turns.

Transformers

A transformer consists of two wire coils linked by a soft magnetic core.

Switch-mode transformers have advantages compared with traditional transformers.

If a transformer is 100% efficient, $V_p I_p = V_s I_s$.

A potential difference is induced in a wire that moves in a magnetic field.

P1 Improve your grade

Page 6

Explain whether a kettle of hot water cools down quicker if its outer surface is coloured white or dark green. **AO2 (3 marks)**

It cools quicker if it is dark green more heat is lost by radiation. (2)

This answer gets 2 out of 3 marks (grade B). The candidate correctly said that heat loss by radiation is affected by the colour of the kettle. For full marks, explain that the heat is lost more quickly from dark surfaces.

Page 7

Explain why convection can take place in a liquid but not in a solid. **AO2 (3 marks)**

In a solid particles are in fixed positions. During convection heat is carried by particles that move.

This answer gets 2 out of 3 marks (grade B). The candidate correctly described the arrangement of particles in a solid, but could also have said that in a liquid they are free to change places. For full marks, explain that particles move from a hotter to a cooler place.

Page 8

Explain why a towel dries quicker on a windy summer day. **AO2 (3 marks)**

Water evaporates from the towel quicker. (1)

This answer gets 1 out of 3 marks (grade C/D). The candidate did not describe how the wind or sun increases evaporation. For full marks, either say wind moves air away from the towel so the air just above the towel does not become saturated or that on a summer's day the air is warmer so particles have more energy so can break bonds more easily.

Page 9

The specific heat capacity of copper is 390 J/Kg °C. Explain whether copper heats up quicker than the same mass of water when they are put in a hot place. The specific heat capacity of water is 4200 J/ Kg °C. **AO2 (3 marks)**

Copper gets hotter than the water.

This answer gets 0 out of 3 marks (grade E/F). Copper heats up quicker but if the water and copper are left for long enough, their final temperature is the same. Remember to include 3 points for a 3 mark question, and to use the information given. For full marks, explain that copper needs less energy than water to heat up by 1 degree. If they absorb the same energy, the copper has a larger increase in temperature.

Page 10

Explain what this Sankey diagram shows in as much detail as possible. **AO2 (3 marks)**

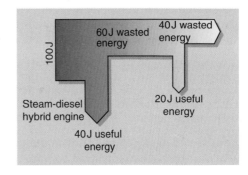

The efficiency is 20%. 60 J of useful energy are transferred and 100 J of wasted energy.

This answer gets 0 marks (grade E/F). The energy is transferred in two stages, which the candidate has confused. For full marks, explain what is meant by efficiency, show calculations and discuss both stages. In the first stage, efficiency is 40%. During the second stage, efficiency is 33%. Overall efficiency is 60%.

Page 11

Two different bulbs are switched on for 10 minutes. Calculate the energy transferred by each one i) a 60 W filament bulb ii) a 10 W energy efficient bulb. **AO2 (4 marks)**

Energy transferred = power x time. The energy transferred is 600 for bulb 1 and 100 for bulb 2.

This answer gets 1 mark (grade D) as the candidate forgot to convert minutes to seconds, and did not show the unit for energy. For full marks, show working. For example the 1st bulb transferred 60 x 10 x 60 = 36 000 Joules.

Page 12

Describe the energy changes taking place in these parts of a coal-fired power station: the burning fuel; the boiler; the turbine; the generator. **AO1 (4 marks)**

burning fuel: chemical → heat

boiler: heat → kinetic

turbine: kinetic (steam) → kinetic (turbine)

generator: kinetic → electrical

This answer gets 4 marks (grade A). The initial energy and final energy had been correctly identified in each case.

Page 13

Explain whether a hydroelectric power station or a coal-fired power station is best for a city located near the coast. **AO3 (5 marks)**

Coal is best (1);

Near the coast rivers flow slowly (1) they cannot be easily dammed (1);

Coal can be transported by water easily (1); it will produce enough electricity for a city (1).

This answer gains 5 marks (grade A)

Page 14

Explain whether a coal-fired power station or a hydroelectric power station is best to cope with surges in demand during the day. **AO3 (3 marks)**

The hydroelectric power station can be started in minutes. However, it causes flooding and can only be used in some parts of the UK. Water can be pumped into the reservoir during quiet periods ready for later use. Coal fired power stations produce greenhouse gases.

This answer gets 2 out of 3 marks (grade B). The comments about environmental issues are irrelevant to the question. To get full marks the candidate should explain that more power stations need to produce electricity if there is a surge in demand for electricity.

Page 15

An echo is a reflected sound wave. Explain why you can hear echoes only in certain places, and why you may hear more than one echo. **AO3 (3 marks)**

Sound waves must bounce off a surface to hear an echo. More than one echo is heard if the sound bounces off several objects.

This answer gets 2 out of 3 (grade B). To get full marks give examples of hard surfaces that sound can echo off e.g. cliffs or buildings.

Page 16

Draw traces to show two sound waves. One sound is lower pitched and twice as loud as the other sound. Label the amplitude and wavelength on each trace. **AO1 (4 marks)**

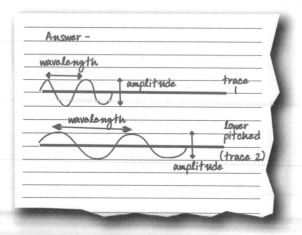

This answer gets 2 out of 4 (grade C/D). The amplitude is wrongly labelled, and the candidate has not shown the trace of a louder note. To get full marks correctly label amplitude as height from mid point to peak, and for a note twice as loud, show a wave with double the amplitude (twice as high).

Page 17

Calculate the speed of radio waves with a wavelength of 10 000 m and frequency of 30 000 Hz. **AO2 (3 marks)**

speed = 30,000÷10,000 = 3 m/s

This answer gets 1 out of 3 marks (grade D). The candidate did not write down the correct equation, or substitute correctly into it. They only get marks for showing the correct unit.

Page 18

Explain which type of electromagnetic wave is the best choice for satellite communications. **AO2 (3 marks)**

Microwaves are best as they are not absorbed by the atmosphere.

This answer gets 2 out of 3 (grade B). To get full marks it is important to show you know why microwaves must travel through the atmosphere, which is that satellites orbit above the earth's atmosphere.

Page 19

Explain two advantages of using space based telescopes. **AO2 (4 marks)**

They can see more as they are closer to the stars, and they can see in all directions.

This answer gets 0 out of 4 marks (grade E/F). The candidate does not realise that the atmosphere has a big impact on the signals reaching the Earth's surface. To get full marks explain that the radiation is not absorbed by the atmosphere so fainter objects can be seen. Another advantage is that objects in space emitting all types of electromagnetic radiation can be detected as the telescope is outside the atmosphere.

Page 20

Explain the evidence we have that supports the Big Bang theory. **AO3 (6 marks)**

The Big Bang was a massive explosion billions of years ago as we can still hear the echo. Galaxies are still moving away from us.

This answer gets 2 out of 6 (grade D/E). Answers like these often allow you to include several points, but expect you to include explanations. The candidate correctly stated the Big Bang theory, but wrongly said the "echo" was heard – it is detected as electromagnetic radiation. To get full marks choose points like these: the red shift shows distant galaxies are moving away from Earth, and further away galaxies move away faster so they used to be closer together.

Background microwave radiation caused by the Big Bang is detected in all directions.

Improve your grade

P2 Improve your grade

Page 22

Sam travels 5 m in 10 s, before stopping for 2 s. He then takes 6 s to return to the start. Draw a distance time graph that shows Sam's journey. **AO2 (3 marks)**

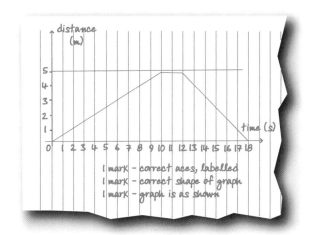

This answer gains 3 out of 3 marks (grade A). The candidate correctly used the information from the question to plot points and then joined the points with a straight line. Distance time graphs like these are one of the few times in physics where a graph "joins the dots".

Page 23

Calculate the total distance travelled by the stone in the first 10 seconds in fig. 3. **AO2 (3 marks)**

distance = area under graph = 80 m

This answer gains 1 out of 3 marks (grade D).

The candidate correctly worked out the area under the graph, but only for the first 4 seconds.

To gain full marks, they should also calculate the distance travelled during 4–10 seconds and add the distances together to get the answer.

Page 24

Describe how drag forces compare with forces from the engine for a car that accelerates, then reaches a steady speed then decelerates. **AO2 (4 marks)**

Drag forces change with speed. When the car accelerates, engine forces are larger than drag forces. At a steady speed there are no drag forces.

This answer gains 2 out of 4 marks (grade C/D).

The candidate forgot to explain what happens when the car decelerates, and was wrong to say there are no drag forces at a steady speed.

To gain full marks, they should say that at a steady speed, forces from engine equal drag forces and as the car decelerates, forces from engine are smaller than drag forces.

Page 25

Sketch the velocity time graph for a marble that is released into a measuring cylinder of water. The marble reaches its terminal velocity. **AO2 (3 marks)**

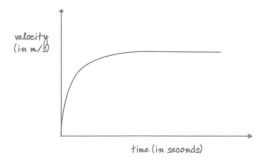

This answer gains 3 out of 3 marks (grade A). The candidate showed the rapid increase in velocity when the drag forces are small, which gets less as the drag forces increase. Eventually there is no increase in velocity when drag forces equals weight. Remember the marble's speed never decreases.

Page 26

a   Calculate the work done lifting a piano 0.8 m into a lorry. The piano weighs 1850 N. **AO2 (3 marks)**

a   work = force x distance = 1850 x 0.8 = 1480

This answer gains 2 out of 3 marks (grade B).

To gain full marks, always include units.

b   Calculate the force needed to drag the piano into the lorry if a ramp 2 m long is used. **AO2 (2 marks)**

b   force = work/distance = 1480/2 = 740 N

This answer gains 2 out of 2 marks (grade A). The candidate included the unit and showed working in this question, which allowed them to gain full marks.

Page 27

Calculate the power of a motor that can lift 600 N in 1 minute. **AO2 (3 marks)**

600W

This answer gets 0 marks. Remember 1 minute is 60 seconds. Always show working. Power = 600/60 = 10 W

Page 28

Two trolleys each with a mass of 1 kg roll towards each other. One trolley travels at 2 m/s to the left and one travels at 3 m/s to the right. They collide and stick together. What is the velocity and direction of the trolleys after the collision? **AO2 (4 marks)**

momentum before = 1 x 2 – 1 x 3 = -1

momentum before = momentum after collision = -1 kg.m/s

speed = 1/1 = 1 m/s

This answer gains 2 out of 4 marks (grade C/D). The candidate did not include a direction for the vehicle after the collision. To gain full marks, a sketch is helpful so you don't miss details – this candidate did not realise the masses stuck together after the collision.

Page 29

Max rubbed a plastic ruler and a metal ruler with a piece of cloth.

a Which ruler became charged? **AO1 (1 mark)**

b Max held the charged ruler near a charged balloon. The balloon moved away from the ruler. Explain why **AO2 (2 marks)**

a the plastic ruler

b the balloon and ruler are charged

This answer gets 2 marks. The plastic ruler gained a charge. The ruler repelled the balloon because they both have the same charge.

Page 30

When the potential difference across a bulb is 6V, the current through it is 0.1A. What is the resistance of the bulb? **AO2 (3 marks)**

Resistance = voltage/current

This answer gets 1 mark. Voltage and potential difference are the same. Resistance = 6V/0.1A = 60 ohms.

Page 31

A series circuit is set up, using four 1.5 V cells in series **AO2 (5 marks)**

a What is the total voltage supplied to the circuit?

a Answer: 6 V (1)

The circuit includes a motor (resistance 10 ohms) and a bulb (resistance 5 ohms).

b What is the total resistance in the circuit?

b Answer: 15 ohm (1)

c Calculate the current in the circuit

c Answer: 6/15 = 0.4 A (1)

d What is the potential difference across each component?

d Answer: 6 V

This answer gains 3 out of 5 marks (grade C). The candidate has worked through the first points correctly, but forgot that in a series circuit the voltage is shared across components.

To gain full marks, include calculations like these:
Bulb: 5 x 0.4 = 2 V Motor: 10 x 0.4 = 4 V

Page 32

Explain why a double insulated appliance does not need an earth wire. **AO1 (2 marks)**

It has a fuse instead

This gets 0 marks. The earth wire stops a metal outer casing from becoming live. Double insulated appliances have a plastic outer casing.

Page 33

Fuses rated at 3 A, 5 A and 13 A are available. Which fuse should be used in each case? **AO2 (3 marks)**

a a lawn mower (power 2000 W)

a 13 A

b a lamp (power 60 W)

b 3 A

c a toaster (power 800 W)

c 5 A

This answer gains 3 out of 3 marks (grade A). The candidate correctly realised that the best fuse has a rating close to, but larger than the current in the equipment. In some questions, you may be asked to show your calculations.

Page 34

Why are the risks from handling an alpha source safely different from the risks from handling a gamma source? **AO3 (3 marks)**

Alpha sources are not dangerous but gamma radiation is. Gamma radiation cannot be used safely

This answer gets 0 marks. Alpha sources are much more dangerous inside the body because they cannot penetrate skin but are very ionising to cells. Gamma radiation is less ionising but penetrates skin and other materials easily. Both should be handled with care.

Page 35

Explain how the Rutherford scattering experiment provided evidence for the nuclear model of the atom. **AO2 (3 marks)**

Most alpha particles were not scattered so the nucleus is small with a positive charge and most of the mass of the atom is in the nucleus.

This answer gains 1 out of 3 marks (grade D). The candidate included correct statements about the nucleus, but did not explain how the evidence leads to these conclusions. To gain full marks, include extra detail such as most alpha particles were not scattered so the nucleus is small. The nucleus must be massive as it could deflect alpha particles. It has a positive charge as it repelled positively charged alpha particles.

Page 36

The half-life of **substance X** is 20 days, and its original activity is 100 000 counts per minute. **Substance Y** has a half-life of 15 days, and an original count rate of 200 000 counts per minute. What is the count rate of each sample after 60 days? **AO2 (3 marks)**

The count rate of X is 100 000/20 which is 5000 and the count rate of Y is 13 333

This answer gains no marks (grade F). The candidate does not know what half-life is (the time taken for the original count rate to halve). For full marks, you should explain that 60 days is 3 half-lives for substance X, so its count rate has halved 3 times and is now an eighth of the original value, i.e. 100 000 / 8 = 12 500

60 days is 4 half lives for substance Y so its count rate is a sixteenth of the original value
i.e. 200 000 / 16 = 12 500

Page 37

Explain why the presence of iron in the Sun is evidence that the Sun formed from the remains of older stars. **AO2 (3 marks)**

Iron is only formed in the later stages of a star's life cycle. The iron spreads through the Universe during a supernova.

This answer gains 2 out of 3 marks (grade B/C) as the candidate did not explain why iron does not form in the Sun. Iron only forms in the final stages of a massive star's lifecycle.

Page 38

Explain how the stages in a star's life cycle are controlled by its mass. **AO2 (3 marks)**

All stars have a main sequence stage. Small stars become red giants and large stars become red supergiants. Small stars do not have a supernova but change into a black dwarf. Large stars become black holes or neutron stars.

This answer gains 2 marks (grade B) It describes the stages but doesn't explain that more massive stars can undergo different fusion reactions.

P3 Improve your grade

Page 40

Explain why ultrasound scans are safer to use than X-rays. **AO2 (2 marks)**

Ultrasound scans are sound waves. They are used to scan pregnant women so must be safe.

This answer gains 0 marks (grade E). It should state that X-rays are ionising, which damages DNA in cells; and ultrasound is not ionising.

Page 41

Explain how a lens can be used as a magnifying glass. **AO2 (3 marks)**

If you look through a convex lens the object appears larger. This magnifies the object.

This answer gains 3 marks (grade A) because it identifies the type of lens, and where it should be placed, and explains what magnification is.

Page 42

Describe the changes taking place in the eye when a person focuses on a nearby object then on a distant object. **AO2 (4 marks)**

The silary muscles push on the lens and push it fatter to change its shape. It has a shorter focal length.

Then the silary muscles pull on the lens and pull it thinner. This means it focuses rays from distant objects on the back of the eye.

This gains 2 marks (grade C). Although a lot of points are correct, there are several mistakes. Remember to learn the spelling of technical words (ciliary muscles) and that muscles cannot push – they can only stretch. The candidate did not mention the role of suspensory ligaments.

Page 43

Explain why double decker buses are less stable than single deck buses. **AO2 (2 marks)**

The centre of mass is higher above the ground for a double decker bus. It is more likely to lie outside the base of the bus.

This answer gains 2 marks (grade A) as it links two ideas - centre of mass is higher for taller objects, and the centre of mass must lie over the base for stability.

Page 44

Calculate the force, F2, in Figure 1 if F1 is 30 N, d1 is 1 m and d2 is 2 m **AO2 (2 marks)**

15

This answer gains 1 mark (grade C). Always remember to include units. The candidate did not show working, so if there was a mistake they would lose all marks.

Page 45

Explain how the centripetal changes in each case:

a   A car slows down as it drives round a bend

b   A bucket of water is swung in circles when it is full of water and when it is empty. **AO1 (2 marks)**

a   The force changes, and becomes stronger.

b   The force is bigger when the bucket is full.

The candidate gains 0 marks – they were asked to explain their answer, but just stated a change.

Part a is wrong (the force decreases because centripetal forces are less when the speed is lower) and part b is correct (the force is greater when the moving mass is larger).

Page 46

Explain how the motor effect is used to make the drum in a washing machine spin. **AO2 (3 marks)**

The motor spins round because it has a coil between magnets. There is a force when a current flows in the coil and this makes the coil spin because of the motor effect.

The candidate gains 3 out of 3 marks (grade A). They should make it clear that the coil is free to spin on an axle, and it is better to define terms clearly (like the motor effect).

Page 47

A transformer has 3000 turns on its primary coil and 6000 turns on its secondary coil.

a   Is it a step-up or a step-down transformer? **AO1 (1 mark)**

b   What is the voltage across the secondary coil if the voltage across the primary coil is 12 V? **AO2 (2 marks)**

a   It is a step up transformer.

b   The voltage is 12 V.

The candidate gains 1 mark for part a and no marks for part b (grade D). You should state and use the turns rule – show working as there are always marks for working even if you make a mistake. Remember to include units!

The correct answer is 12 V × 6000 / 3000 = 24 V

Understanding the scientific process

As part of your assessment, you will need to show that you have an understanding of the scientific process – How Science Works.

This involves examining how scientific data is collected and analysed. You will need to evaluate the data by providing evidence to test ideas and develop theories. Some explanations are developed using scientific theories, models and ideas. You should be aware that there are some questions that science cannot answer and some that science cannot address.

Collecting and evaluating data

You should be able to devise a plan that will answer a scientific question or solve a scientific problem. In doing so, you will need to collect data from both primary and secondary sources. Primary data will come from your own findings – often from an experimental procedure or investigation. While working with primary data, you will need to show that you can work safely and accurately, not only on your own but also with others.

Secondary data is found by research, often using ICT – but do not forget books, journals, magazines and newspapers are also sources. The data you collect will need to be evaluated for its validity and reliability as evidence.

Presenting information

You should be able to present your information in an appropriate, scientific manner. This may involve the use of mathematical language as well as using the correct scientific terminology and conventions. You should be able to develop an argument and come to a conclusion based on recall and analysis of scientific information. It is important to use both quantitative and qualitative arguments.

Changing ideas and explanations

Many of today's scientific and technological developments have both benefits and risks. The decisions that scientists make will almost certainly raise ethical, environmental, social or economic questions. Scientific ideas and explanations change as time passes and the standards and values of society change. It is the job of scientists to validate these changing ideas.

How science ideas change

From the information you have learnt, you will know that science is a process of developing, then testing theories and models. Scientists have been carrying out this work for many centuries and it is the results of their ideas and trials that has provided us with the knowledge we have today.

However, in the process of developing this knowledge, many ideas were put forward that seem quite absurd to us today.

> In 1692, the British astronomer Edmund Halley (after whom Halley's Comet was named) suggested that the Earth consisted of four concentric spheres. He was trying to explain the magnetic field that surrounds the Earth and suggested that there was a shell of about 500 miles thick, two inner concentric shells and an inner core. Halley believed that these shells were separated by atmospheres, and each shell had magnetic poles with the spheres rotating at different speeds. The theory was an attempt to explain why unusual compass readings occurred. He also believed that each of these inner spheres, which was constantly lit by a luminous atmosphere, supported life.

Reliability of information

It is important to be able to spot when data or information is presented accurately, and just because you see something online or in a newspaper does not mean that it is accurate or true.

Think about what is wrong in this example from an online shopping catalogue. Look at the answer at the bottom of the page to check that your observations are correct.

FROM BOX TO AIR IN UNDER TWO MINUTES!

Simply unroll the airship and, as the black surface attracts heat, watch it magically inflate.

Seal one end with the cord provided and fly your 8-metre, sausage-shaped kite.

✔ Good for all year round use.

✔ Folds away into box provided.

✔ A unique product – not for the faint hearted.

✔ Educational as well as fun!

Once the airship is filled with air, it is warmed by the heat of the sun.

The warm air inside the airship makes it float, like a full-sized hot-air balloon.

Answer
Black absorbs heat, it does not attract it.

Glossary / Index

The glossary contains terms useful for your revision. Page numbers are given for items that are covered in this book.

crumple zone 26 part of a vehicle designed to absorb energy in an accident, so reducing injuries to passengers

current 11, 14, 29–31, 33, 39, 46 flow of electricity around a circuit – carried by electrons through solids and by ions through liquids

D

deceleration see negative acceleration

diffraction 15, 17, 21 change in the direction of a wave caused by passing through a narrow gap or round an obstacle such as a sharp corner

digital signal 19 communications signal sent as an electromagnetic wave that is switched on and off very rapidly

diode 30 semiconductor device that allows an electric current to flow through it in only one direction

direct current (d.c.) 32, 39 electric current where the direction of the flow of current stays constant, as in cells and batteries

distance–time graph 22 graph showing how the distance an object travels varies with time: its gradient shows speed

distribution 21 the transmission of electricity from a power station to homes and businesses

DNA 18 deoxyribonucleic acid – the chemical from which chromosomes are made: its sequence determines genetic characteristics, such as eye colour

Doppler effect 20 change in wavelength and frequency of a wave that an observer notices if the wave source is moving towards them or away from them

E

earth wire 32 wire connecting the case of an electrical appliance, through the earth pin on a three-pin plug, to earth

earthed 32 safety feature where part of an appliance is connected to earth to protect users from electrocution if there is a fault

earthquake 15, 17 shaking and vibration at the surface of the Earth resulting from underground movement or from volcanic activity

echo 16 reflection of a sound wave

efficiency 10, 12, 21 a measure of how effectively an appliance transfers the energy that flows into the appliance into useful effects

elastic collision 28 collision where colliding particles or objects bounce apart after collision

elastic potential energy 25 energy stored in an object because it is stretched, compressed or deformed, and released when the object returns to its original shape

elastic 25, 39 material that returns to its original shape when the force deforming it is removed

electrical power 11–14, 21 a measure of the amount of energy supplied each second

electricity generator 12–13 device for generating electricity

electromagnetic (EM) radiation 17–18, 20, 34 energy transferred as electromagnetic waves

electromagnetic spectrum 40 electromagnetic waves ordered according to wavelength and frequency – ranging from radio waves to gamma rays

electromagnetic waves 15, 17–19, 21 a group of waves that transfer energy – they can travel through a vacuum and travel at the speed of light

electron 7, 11, 28, 30, 34 small particle within an atom that orbits the nucleus (it has a negative charge)

electrostatic induction 29 electric charge induced on an object made of an electrical insulator, by another electrically charged object nearby

emit 21 an object emits energy when energy is transferred away from the object as infrared radiation, decreasing the temperature of the object

energy output 6–7, 9–11, 15, 21, 26 the energy transferred away from a device or appliance – it can be either useful or wasted

energy 6–8, 21, 29 the ability to 'do work'

environment 8, 13 an organism's surroundings

evaporation 6, 8, 21 change of state where a substance changes from liquid to gas at a temperature below its boiling point

F

fibre optic cable 19 glass fibre that is used to transfer communications signals as light or infrared radiation

filament bulb 11, 30 lightbulb giving out light by current flowing through a fine wire and heating the wire until it glows white hot

fossil fuel 12–14 fuel such as coal, oil or natural gas, formed millions of years ago from dead plants and animals

freezing 7 change of state in which a substance changes from a liquid to a solid

frequency 16–17, 20–21, 32 the number of waves passing a set point per second

friction 26, 45   force acting at points of contact between objects moving over each other, to resist the movement

fuse 29, 32–33, 39   a fine wire that melts if too much current flows through it, breaking the circuit and so switching off the current

G

gamma rays 17–19, 34, 36, 39   ionising electromagnetic radiation – radioactive and dangerous to human health

geothermal power station 13   power station generating electricity using the heat in underground rocks to heat water

global warming 6, 13   gradual increase in the average temperature of Earth's surface

gravitational potential energy 27, 39, 45   energy that an object has because of its position, for example, increasing the height of an object above the ground increases its gravitational potential energy

gravity 20, 27, 45   the attractive force acting between all objects with mass – on Earth the attractive force due to gravity pulls objects downwards

H

Hooke's Law 25   for an elastic object, the extension is proportional to the force applied, provided the limit of proportionality is not exceeded

hydroelectric power station 13–14, 21   power station generating electricity using the energy from water flowing downhill

I

inelastic collision 28   collision where the colliding particles or objects stick together after collision

infrared radiation 6, 8, 17–19   energy transferred as heat – a type of electromagnetic radiation

ion 34, 39   atom (or group of atoms) with a positive or negative charge, caused by losing or gaining electrons

ionise 10   to cause electrons to split away from their atoms (some forms of EM radiation are harmful to living cells because they cause ionisation)

isotopes 34, 36, 39   forms of element where their atoms have the same number of protons but different numbers of neutrons

J

joule 6, 11   unit used to measure energy

K

kilowatt-hour 11–12   the energy transferred in 1 hour by an appliance with a power rating of 1 kW (sometimes called a 'unit' of electricity)

kinetic energy 7, 12, 26–27, 29, 39, 45   energy an object has because of its movement – it is greater for objects with greater mass or higher speed

kinetic theory 21   model used to explain how energy is transferred by particles in a substance

L

LDR (light dependent resistor) 29–30   resistor with a resistance that decreases when light is shone on it

LED (light emitting diode) 11, 30   diode that gives off light when a current flows through it

longitudinal wave 15–17, 21   a wave in which the direction that the particles are vibrating is the same as the direction in which the energy is being transferred by the wave

M

main sequence star 38   a star in which nuclear fusion reactions combine small atomic nuclei into elements with larger nuclei

mains supply 47   domestic electricity supply – in the UK, mains supply is 230 V at 50 Hz

mass number 34   total number of protons and neutrons in the nucleus of an atom – always a whole number

mass 39, 43, 49   a measure of the amount of 'stuff' in an object

mechanical wave 15, 17   wave in which energy is transferred by particles or objects moving, such as a wave on a string or a water wave

melting 7   change of state of a substance from liquid to solid

microwaves 17–18, 21   non-ionising radiation – used in telecommunications and in microwave ovens

momentum 28, 39   mass of a moving object multiplied by its velocity – a vector quantity having both size and direction

mutation 18   a change in the DNA in a cell

renewable resource 13   energy resource that is constantly available or can be replaced as it is used

resistance (electrical) 30, 39   measure of how hard or how easy it is for an electric current to flow through a component

resultant force 23–24, 39   the single force that would have the same effect on an object, as all the combined forces that are acting on the object

S

Sankey diagram 10   diagram showing how the energy supplied to something is transferred into 'useful' or 'wasted' energy

series circuit 29, 31, 39   electrical circuit with only one possible path for the current to flow around

solar panel 21   panel that uses the Sun's energy to heat water

solar power station 13–14   power station generating electricity using energy transferred by the Sun's radiation

specific heat capacity 9, 21   a measure of the amount of energy needed to raise the temperature of 1 kg of a substance by 1 °C

speed of light 17   speed at which electromagnetic radiation travels through a vacuum – 300 000 000 metres per second

speed 22   how quickly an object is moving, usually measured in metres per second (m/s)

states of matter 7, 21   substances can exist in three states of matter (solid, liquid or gas) – changes from one state to another are called changes of state

static electricity 28, 39   an electric charge on an insulating material, caused by electrons flowing onto or away from the object

step-down transformer 14, 28   transformer that changes alternating current to a lower voltage

step-up transformer 12, 14, 48   transformer that changes alternating current to a higher voltage

stopping distance 24, 39   total distance it takes a vehicle to stop – the sum of thinking distance and braking distance

sub-atomic particle 34   particle that make up an atom, such as proton, neutron or electron

surface area (of a solid reactant) 8, 21   measure of the area of an object that is in direct contact with its surroundings

T

telecommunications 18   communications over long distances using various types of electromagnetic radiation

terminal velocity 25, 39   maximum velocity an object can travel at – at terminal velocity, forward and backward forces are the same

thermistor 29–30   resistor made from semiconductor material: its resistance decreases as temperature increases

thinking distance 24, 39   distance a vehicle travels while a signal travels from the driver's eye to brain and then to foot on the brake pedal: thinking distance increases with vehicle speed

three core cable 32   electrical cable containing three wires, live, neutral and earth

three-pin plug 32   type of plug used for connecting to the mains supply in the UK: it has three pins, live, neutral and earth

tidal power station 13, 21   power station generating electricity using the energy transferred by moving tides

tracer 36   radioactive element used to track the movement of materials, such as water through a pipe or blood through organs of the body

transformer 47–48   device by which alternating current of one voltage is changed to another voltage

transverse wave 15, 17, 21   a wave in which the vibration of particles is at right angles to the direction in which the wave transfers energy

turbine 13–14   device for generating electricity – the turbine has coils of wire that rotate in a magnetic field to generate electricity

U

ultraviolet radiation 17–18   electromagnetic radiation that can damage human skin

upright image 16, 42, 48   image that is the same way up as the object

U value 21   a measure of how easily energy is transferred through a material as heat

V

vacuum 17, 21   a space in which there are no particles of any kind

velocity 17, 22, 39, 45   measure of how fast an object is moving in a particular direction

velocity–time graph 22, 25   graph showing how the velocity of an object varies with time: its gradient shows acceleration

virtual image 16, 42, 48   image that can be seen but cannot be projected onto a screen (a mirror forms a virtual image behind the mirror)

voltage 11, 14, 30–31   a measure of the energy carried by an electric current (the old name for potential difference)

W

wasted energy 10, 12   energy that is transferred by a device or appliance in ways that are not wanted, or useful

watt 11, 14, 27   unit of energy transfer – one watt is a rate of energy transfer of one joule per second

wave equation 17   the speed of a wave is always equal to its frequency multiplied by its wavelength

wave power 13–14, 21   electricity generation using the energy transferred by water waves as the water surface moves up and down

wavelength 15, 17, 20   distance between two wave peaks

weight 24   the downward force on a mass due to gravity, measured in newtons (N)

wind turbine 13   device generating electricity by using the energy in moving air to turn a turbine and a generator

work 26   amount of energy transferred to an object by a force moving the object through a distance:
work done = force × distance moved in the direction of the force

X

X-rays 18–19, 40, 48   ionising electromagnetic radiation – used in X-ray photography to generate pictures of bones

Collins

Workbook

NEW GCSE SCIENCE

Physics

for AQA A Higher

Author: Nicky Thomas

Revision guide +
Exam practice workbook

The key to successful revision is finding the method that suits you best. There is no right or wrong way to do it.

Before you begin, it is important to plan your revision carefully. If you have allocated enough time in advance, you can walk into the exam with confidence, knowing that you are fully prepared.

Start well before the date of the exam, not the day before!

It is worth preparing a revision timetable and trying to stick to it. Use it during the lead up to the exams and between each exam. Make sure you plan some time off too.

Different people revise in different ways and you will soon discover what works best for you.

Some general points to think about when revising

- Find a quiet and comfortable space at home where you won't be disturbed. You will find you achieve more if the room is ventilated and has plenty of light.

- Take regular breaks. Some evidence suggests that revision is most effective when tackled in 30 to 40 minute slots. If you get bogged down at any point, take a break and go back to it later when you are feeling fresh. Try not to revise when you're feeling tired. If you do feel tired, take a break.

- Use your school notes, textbook and this Revision guide.

- Spend some time working through past papers to familiarise yourself with the exam format.

- Produce your own summaries of each module and then look at the summaries in this Revision guide at the end of each module.

- Draw mind maps covering the key information on each topic or module.

- Review the Grade booster checklists on pages 109–110.

- Set up revision cards containing condensed versions of your notes.

- Prioritise your revision of topics. You may want to leave more time to revise the topics you find most difficult.

Workbook

The Workbook allows you to work at your own pace on some typical exam-style questions. You will find that the actual GCSE questions are more likely to test knowledge and understanding across topics. However, the aim of the Revision guide and Workbook is to guide you through each topic so that you can identify your areas of strength and weakness.

The Workbook also contains example questions that require longer answers (**Extended response questions**). You will find one question that is similar to these in each section of your written exam papers. The quality of your written communication will be assessed when you answer these questions in the exam, so practise writing longer answers, using sentences. The **Answers** to all the questions in the Workbook are detachable for flexible practice and can be found on pages 113–125.

At the end of the Workbook there is a series of **Revision checklists** that you can use to tick off the topics when you are confident about them and understand certain key ideas.

Remember

There is a difference between learning and revising.

When you revise, you are looking again at something you have already learned. Revising is a process that helps you to remember this information more clearly.

Learning is about finding out and understanding new information.

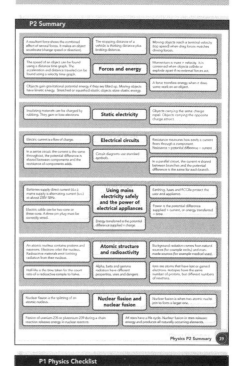

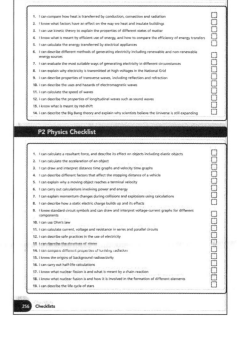

Energy

1 (a) Describe the energy transfers taking place when a hairdryer is turned on.

_____

_____ [4 marks]

(b) Where is the energy finally transferred to? _____ [1 mark]

(c) A student investigates how quickly hairdryers transfer energy by timing how long it took for each hairdryer to dry a piece of damp cloth. Here are their results:

Model of hairdryer	Time needed to dry damp cloth
Traveller	2 minutes 30 seconds
Olympus1000	1 minute 20 seconds
Hairdry	2 minutes 10 seconds

Explain which hairdryer transferred energy quickest.

_____ [2 marks]

(d) State two variables that the student should control during this investigation.

_____ [2 marks]

2 A company has produced a battery-operated hairdryer which uses rechargeable batteries. Explain whether this hairdryer wastes more energy than a hairdryer operated from the mains supply.

_____

_____ [4 marks]

Infrared radiation

1 Solar panels are fitted on house roofs and are used to heat water.
(a) Explain which is the best colour for the manufacturer to choose for the solar panels

_____

_____ [3 marks)

(b) Why are solar panels normally fitted to south-facing roofs?

_____ [2 marks]

(c) Solar panels should be insulated to avoid heat losses at night or during winter. Explain why.

_____ [3 marks]

2 A student uses a probe to measure the amount of infrared radiation emitted by a beaker of water at room temperature.
(a) Explain whether the probe registers infrared emissions from the beaker.

_____ [2 marks]

(b) She places the beaker of water in a fridge. Explain how this affects the amount of infrared radiation emitted and absorbed.

_____

_____ [4 marks]

Kinetic theory

1 (a) Explain what melting means.

_____ [1 mark]

(b) A piece of ice left on a plate starts to melt. Explain the process of melting by writing about particles.

_____

_____ [4 marks]

(c) Why do the plate and its surroundings cool down when ice melts?

_____ [2 marks]

(d) Explain why the ice melts more quickly in a warmer room.

_____ [1 mark]

D–C

2 Explain what affects the energy of particles by comparing the particles in a piece of copper and particles in the air which are at the same temperature.

_____

_____ [4 marks]

B–A*

Conduction and convection

1 (a) Explain why convection cannot take place in solids.

_____ [2 marks]

(b) Use your ideas about heat transfers to explain how the thermos flask keeps a drink hot.

_____

_____

_____

_____

_____

_____

[6 marks]

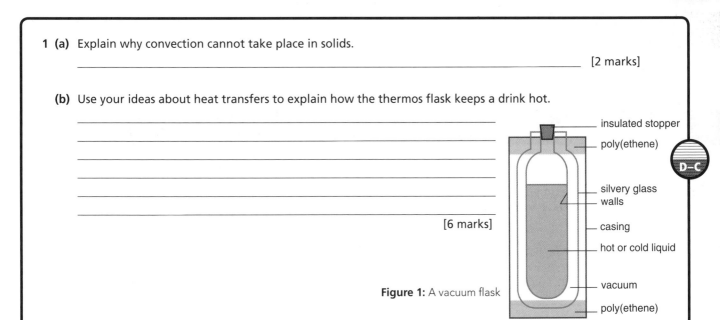

insulated stopper

poly(ethene)

D–C

silvery glass walls

casing

hot or cold liquid

vacuum

poly(ethene)

Figure 1: A vacuum flask

2 Computers can get extremely hot, possibly damaging components inside them. A heat sink is used to cool the computer by transferring heat to the surroundings. Explain why a heat sink in a computer is made from metal rather than plastic.

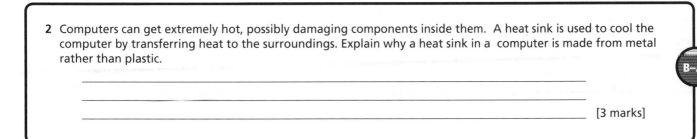

_____

_____

_____ [3 marks]

B–A*

Evaporation and condensation

1 Suna hangs her washing on a washing line outside.

   (a) What is meant by evaporation?

   _____ [1 mark]

   (b) Write down two differences between evaporation and boiling.

   _____ [2 marks]

   (c) Write down two weather conditions that make evaporation take place more quickly. Give a reason each weather condition affects the rate of evaporation

   (i) Condition 1_____
   Why it increases the rate of evaporation

   _____ [2 marks]

   (ii) Condition 2_____
   Why it increases the rate of evaporation

   _____ [2 marks]

D–C

2 Many men use aftershave, which contains alcohol. Alcohol evaporates quickly, cooling their skin. Peter is investigating the hypothesis that aftershave will cool a cotton wool pad more quickly than water does Write down a plan for an experiment to test this hypothesis. Include a list of equipment and a method.

   _____
   _____
   _____
   _____
   _____
   _____ [6 marks]

B–A*

Rate of energy transfer

1 When a potato is cut into smaller pieces, it cooks more quickly. Explain why this happens using these terms in your answer: surface area; conduction; radiation.

   _____
   _____
   _____
   _____ [4 marks]

D–C

2 The boiling temperature of water is 100 °C, and the boiling temperature of cooking oil is 160 °C Use your ideas about heat transfers to explain why chips cook more quickly when they are fried rather than boiled.

   _____
   _____ [2 marks]

B–A*

Insulating buildings

D–C

1 A homeowner paid for an energy survey and report for their home. The report suggested several ways to reduce energy losses in the home.

(a) Suggest one reason why the homeowner was prepared to pay for a report.

_____ [1 mark]

Here are four suggestions made by the consultant.

Change	Estimated Cost	Estimated annual Saving
Change all light bulbs to energy efficient bulbs	£80	£95
Increase loft insulation to a depth of 15 cm	£200	£120
Set central heating thermostat on to a lower temperature setting	£0	£40
Put the central heating onto an automatic timer switch	£15	£20

All of these measures are put in place in the first year.

(b) Write down the overall savings over five years.

_____ [3 marks]

(c) Loft insulation reduces heat losses because it slows down convection and conduction. Explain whether it has a high or low U-value.

_____ [2 marks]

(d) State two other methods to reduce heat losses in the home. Explain how each method reduces heat losses.

_____
_____ [4 marks]

2 A local council provides grants for people to install energy saving measures. These measures do not include double-glazing, partly because of the high installation costs and the small average annual energy savings. Explain whether you agree with the council's decision.

B–A*

_____
_____ [5 marks]

Specific heat capacity

1 (a) Explain what is meant by specific heat capacity.

_____ [1 mark]

D–C

(b) The specific heat capacity of aluminium is 900 J/kg °C. Calculate the temperature rise when a 250 g block of aluminium is supplied with 1800 J of energy. Write down the equation you use, then clearly show how you calculate your answer.

_____ [3 marks]

2 Use your ideas of specific heat capacity to explain whether it would be better to use copper or aluminium for a saucepan base.
The specific heat capacity of aluminium is 900 J/kg °C, and the specific heat capacity of copper is 385 J/kg °C.

B–A*

_____ [3 marks]

Energy transfer and waste

1 The picture shows a Sankey diagram for a petrol engine.

   (a) Use the diagram to write down the energy equation for the petrol engine.

   _____

                                     [2 marks]

   (b) The energy transfer takes place in two stages in the car. State how most of the energy wastage occurs.

   _____

                                     [1 mark]

   (c) Describe how conservation of energy is shown on a Sankey diagram.

   _____ **[3 marks]**

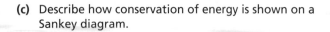

Figure 1: Sankey diagram for a petrol engine

(Diagram labels: 1000 J chemical energy input; 850 J kinetic energy in engine; 300 J useful kinetic energy in car; 150 J wasted heat in exhaust gases; 550 J wasted heat energy in moving engine parts)

---

2 An improved engine has been designed which changes 1000 J of chemical energy into 500 J of kinetic energy. The rest is wasted heat. Draw a Sankey diagram to show this transfer

                                     [4 marks]

Efficiency

1 The diagram shows the energy transfer for an electric fan.

   (a) Use the Sankey diagram to calculate the efficiency of the electric fan.

   _____

                                     [2 marks]

   (b) Explain whether all new electrical appliances should display Sankey diagrams for people to compare their energy efficiency.

   _____

   _____

   _____

                 [3 marks]

Figure 1: Sankey diagram for an electric fan

(Diagram labels: 50 J energy input; 30 J useful energy transferred to spin fan; 5 J wasted energy as sound; 15 J wasted energy as heat)

---

2 A manufacturer claims that a new design of battery-operated motor is 99%. Explain why this claim needs to be checked carefully.

   _____

   _____

   _____ **[3 marks]**

Electrical appliances

1 A company has designed a wind-up mobile phone battery recharger.

(a) Describe the useful energy changes that take place when a person winds up the recharger and charges up the battery.

_____ -> _____ -> _____   [3 marks]

(b) Explain one advantage of;

(i) using a wind-up battery recharger

_____

_____   [2 marks]

(ii) using a battery in the mobile phone.

_____

_____   [2 marks]

D–C

2 Lighting in gardens is becoming more popular. There are many reasons for this, for example improving safety, security and the appearance of the garden at night. Increasingly, these lighting systems do not use mains electricity.

(a) Suggest two reasons why manufacturers have been developing an alternative to mains electricity.

_____

_____   [2 marks]

B–A*

(b) Explain why many people chose solar panels instead of batteries for providing energy for garden lights.

_____

_____   [2 marks]

Energy and appliances

1 Calculate the energy transferred when an 850 W microwave oven is switched on for 15 minutes.

_____

_____

_____   [3 marks]

D–C

2 The flex for a microwave oven is thinner than the flex fitted to an electric oven. Explain why this is a safety feature.

_____

_____

_____

_____   [4 marks]

B–A*

The cost of electricity

1

	Filament bulb	Energy efficient bulb
Power	100 W	16 W
Cost when new	30p	£5.00
Lifetime	1 year	5 years

(a) Each bulb is used for 1500 hours per year. Use this information to calculate the energy used by the filament bulb in a year in kilowatt-hours.

_____ kWh   [2 marks]

(b) If each kilowatt-hour costs 15p, calculate the total cost of using the filament bulb over five years.

_____

_____   [3 marks]

(c) The equivalent cost if energy efficient bulbs are used is less than £19. Explain why the use of energy efficient bulbs is being encouraged for environmental reasons.

_____

_____   [2 marks]

D–C

2 Many environmental organisations encourage people to turn equipment off at the plug rather than leaving it on standby. Evaluate the benefits and disadvantages of manufacturing equipment to automatically turn off rather than to remain on standby.

_____

_____

_____   [6 marks]

B–A*

Power stations

1 (a) Describe the useful energy transfers taking place in a gas-fired power station.

_____

_____

_____   [4 marks]

D–C

(b) Explain why a combined heat and power station is more efficient than a power station that just generated electricity.

_____

_____

_____   [4 marks]

2 Many power stations use technology developed over 20 years ago, and are not as efficient as modern power stations. Explain whether we should convert existing older power stations to become more efficient, or whether we should develop new technologies to use in new power stations.

_____

_____

_____   [4 marks]

B–A*

Renewable energy

1 Both Iceland and Norway use renewable sources of energy to generate electricity. Iceland is a volcanic island and Norway is a mountainous country with many valleys and fast flowing rivers. Suggest which is the most suitable renewable energy resource for each country to use to generate electricity, explaining your answers.

_____

_____

_____

_____ [4 marks]

D–C

2 Many wind farms are being built in the UK. These include onshore and offshore wind farms. Critics say that wind farms do not generate electricity when it is calm or stormy, so it is not practical to depend on them. Supporters say that giant batteries under the turbines could store energy generated by wind turbines for future use. Outline one advantage and one disadvantage of using giant batteries in wind farms for the UK.

_____

_____ [2 marks]

B–A*

Electricity and the environment

1 There are plans to build a new power station near a large city. The choice is between a coal-fired power station or wind turbines set on nearby hills in a local beauty spot.
Give one environmental advantage and one environmental disadvantage for each of these schemes.

Type of station	Advantage of the scheme	Disadvantage of the scheme
Coal-fired		
Wind turbines		

[4 marks]

D–C

2 One green campaigner feels that renewable energy does not harm the environment. Explain two reasons that show how using biofuels can damage the local environment.

_____

_____

_____

_____ [4 marks]

B–A*

Making comparisons

1 (a) Explain two factors that may affect the demand for electricity during the day.

_____

_____

_____ [3 marks]

D–C

(b) A reliable energy source can be used to generate electricity whenever it is required.
Put these energy sources in order of reliability, starting with the least reliable first.
gas, wind, waves, tidal

_____ [3 marks]

2 Critics of wind turbines say that even with large numbers of wind farms, new fossil-fuel power stations are needed to supply electricity when the weather is unsuitable. Supporters say that it is possible to predict when the weather means wind turbines cannot function. This causes less disruption to the supply of energy than an unexpected failure of a single very large fossil-fuel power station.
Use these ideas to discuss whether large numbers of wind farms are the best solution to the UK's energy problems.

B–A*

_____

_____

_____

_____

_____ [5 marks]

The National Grid

1 The National Grid distributes electricity throughout the UK. Overhead high voltage transmission lines transmit electricity across the country at voltages up to 400 000 V. Distribution lines transmit electricity at voltages up to 33 000 V throughout an area.

(a) Why is less energy wasted when electricity is transmitted at higher voltages?

_____

_____ [2 marks]

D–C

(b) Calculate the current in a high voltage transmission line that carries 200 000 000 W.

_____

_____

_____ [3 marks]

2 Pylons many metres above the ground carry high voltage transmission lines. This reduces risk of damage but causes visual pollution. Many people feel that underground power cables would be a better solution although this is more expensive, and still involves damage to the surroundings.
Explain whether power lines should be carried on pylons or should be buried underground.

B–A*

_____

_____

_____ [4 marks]

What are waves?

1 Describe two differences and two similarities between transverse waves and longitudinal waves.

Difference 1 _____

Difference 2 _____

Similarity 1 _____

Similarity 2 _____ [4 marks]

D–C

2 Earthquakes cause seismic waves. P-waves are longitudinal waves and s-waves are transverse waves.

  (a) Seismic waves are mechanical waves. What is meant by a mechanical wave?

_____ [1 mark]

  (b) Explain why the same earthquake is detected in different places at different times.

_____

_____

_____

_____

_____

_____

_____ [4 marks]

B–A*

Changing direction

1 Two people had a conversation in a room. The door was open and the conversation could be clearly heard on the other side of the wall. Use your ideas of diffraction to explain why.

_____

_____

_____

_____

_____ [3 marks]

D–C

2 Use your ideas about refraction to explain why a swimming pool looks shallower than it really is.

_____

_____

_____

_____

_____ [3 marks]

B–A*

Sound

1 (a) A musician plays a note with a frequency of 440 Hz. Another musician plays a slightly higher pitched note. Write down a possible frequency of this note.

_____ [1 mark]

(b) How does the wavelength of a sound change when it becomes higher pitched?

_____ [1 mark]

(c) What changes when a sound becomes louder?

_____
_____ [2 marks]

D–C

2 The picture shows a trace of a sound from an oscilloscope.

(a) Draw the trace that would be seen if a quieter higher pitched sound was heard. [2 marks]

(b) A sound engineer is making a recording of a band, and will mix the sounds later to make the final version. Explain how oscilloscopes can help the sound engineer.

_____
_____
_____

[3 marks]

B–A*

Light and mirrors

1 Describe how the image of a child in a plane mirror compares with the real child by choosing the correct words.

The image is *upright/inverted.*
It is *real/virtual.*
It is the *same distance behind/further behind* the mirror than the object is in front. [3 marks]

D–C

2 The diagram shows a person looking at some writing in a mirror.
Complete the ray diagram to show why the object is laterally inverted.

abcdef

[4 marks]

B–A*

Using waves

1 Match the type of electromagnetic wave with these uses.

Type of electromagnetic radiation	Use
Radio wave	Photography
Microwave	Remote control
Infrared radiation	BBC TV broadcasts
Visible light	Satellite TV

[4 marks]

2 Mobile phone coverage in many parts of the country can be more unreliable than radio reception. Explain why?

_____

_____

_____

_____

_____ [5 marks]

The electromagnetic spectrum

1 Radio waves can be detected from galaxies, such as the Milky Way. If the frequency of radio waves is measured as 2 700 000 kHz, calculate the wavelength of the radiation. Electromagnetic waves travel at 300 000 km/s

_____

_____

_____

_____ [4 marks]

2 Select the correct answer from the following sentence.
The electromagnetic spectrum is a continuous spectrum of waves whose wavelength varies from more than *1000 / 10 000 / 1 000 000 m to 10^{-7} /10^{-10} /10^{-15} m*

_____

_____ [2 marks]

Dangers of electromagnetic radiation

1 Rachel put sun block on her skin before she went out in the Sun.

(a) What type of electromagnetic radiation does sun block protect her from?

_____ [1 mark]

(b) Write down three forms of harm caused by too much exposure to this type of radiation.

_____

_____

_____ [3 marks]

D–C

(c) Write down two other ways that Rachel could reduce her exposure to radiation from the Sun.

_____

_____

_____

_____ [4 marks]

---

2 (a) What effect does shorter wavelength electromagnetic radiation have on cells when absorbed by molecules within the cells?

_____ [1 mark]

B–A*

(b) A patient is undergoing radiotherapy as part of a treatment for cancer. Explain why carefully targeted radiotherapy is a useful treatment for cancer.

_____

_____

_____ [3 marks]

Telecommunications

1 The Olympic Games brings together sportsmen and women from all over the world. Events from the Olympics are transmitted to audiences worldwide. Explain how electromagnetic waves can be used to allow viewers to see live transmissions of the London Olympics in Australia.

D–C

_____

_____

_____ [3 marks]

2 Sky satellite dishes are fixed in one position to receive signals from a communications satellite. The satellite is orbiting the Earth constantly above the equator.
Explain why the signals from the satellite can be received at all times by the dish.

B–A*

_____

_____

_____ [3 marks]

Cable and digital

1 (a) What is the name given to the type of signal shown in figure A?

_____

[1 mark]

(b) Explain why the quality of these signals can get worse when they are transmitted long distances through copper cables.

_____

_____

_____

_____

[2 marks]

A

Searching space

1 (a) Use the table to explain why we need different telescopes to detect different objects in space.

_____

_____

_____

_____

[2 marks]

Radiation	Objects 'seen' in space
gamma ray	neutron stars
X-ray	neutron stars
ultraviolet	hot stars, quasars
visible	stars
infrared	red giants
far infrared	protostars, planets
radio	pulsars

(b) Telescopes that detect radiation with long wavelengths need large receiving dishes because the radio signals from objects in space are very weak. Explain whether a radio telescope is larger or smaller than an optical telescope that detects visible light.

_____

_____

_____

_____

[2 marks]

Waves and movement

1 (a) When a car passes a person, its sound appears to change pitch. What is the name of this effect?

_____ [1 mark]

D–C

(b) The diagram shows sound waves coming from a car. On the diagram, show how the sound wave would appear if the car was travelling away from the person.

[2 marks]

2 The Universe is expanding and galaxies in the Universe are moving apart at different speeds. We can see this because of the red shift.

(a) What is meant by the red shift?

B–A*

_____

_____ [2 marks]

(b) Explain how the red shift shows that galaxies are moving apart at different speeds.

_____

_____ [2 marks]

Origins of the Universe

1 Scientists have developed several models of the start of the Universe. One model is the Steady State theory. This theory describes a universe that is constantly expanding. Matter is being constantly created. The Big Bang theory is another theory that is more widely accepted.

(a) Describe the Big Bang theory of the creation of the Universe.

_____

_____

D–C

_____ [3 marks]

(b) Describe the evidence that supports the Big Bang theory.

_____

_____

_____ [3 marks]

2 Explain why cosmic microwave background radiation is such strong evidence for the Big Bang.

B–A*

_____

_____ [3 marks]

Extended response question

Use your knowledge of the properties of electromagnetic waves to explain why different members of the electromagnetic spectrum are used for different sorts of communication.

The quality of written communication will be assessed in your answer to this question.

_____

_____

_____

_____

_____

_____

_____

_____

_____

_____

_____

_____

_____

_____

_____

_____

_____

_____

_____

_____

_____

_____

_____

_____

_____

_____

_____

_____

_____

_____

_____

_____

_____

_____

_____

_____

_____

_____

_____

_____

_____

_____

_____

_____

_____

_____

_____

[6 marks]

See how it moves

1 The distance–time graph shows the motion of a bus.

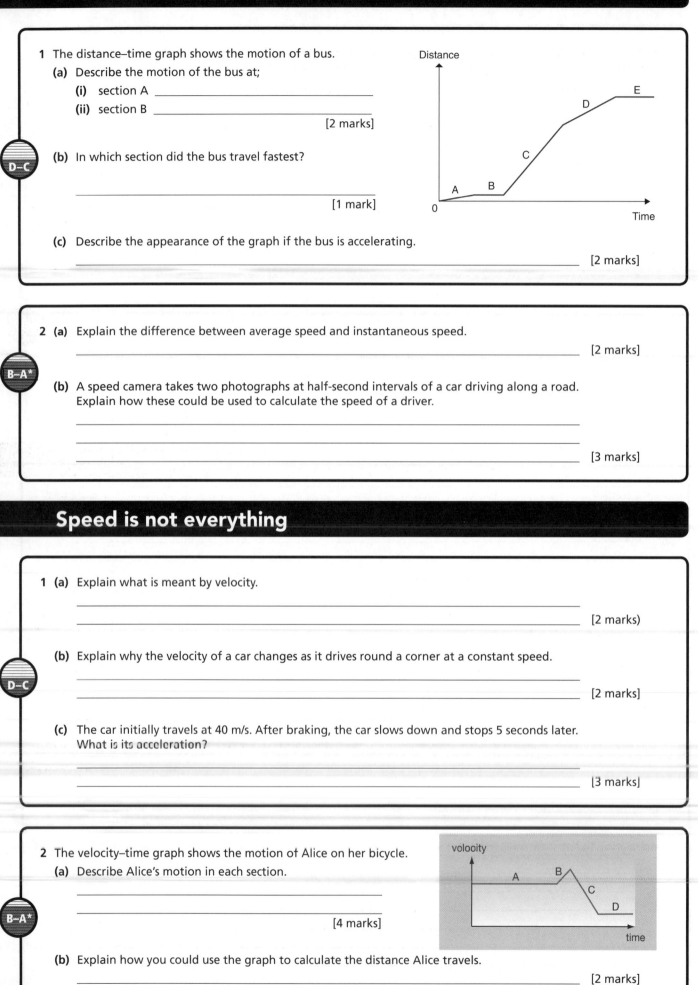

(a) Describe the motion of the bus at;

(i) section A _____

(ii) section B _____

[2 marks]

(b) In which section did the bus travel fastest?

_____

[1 mark]

(c) Describe the appearance of the graph if the bus is accelerating.

_____ [2 marks]

2 (a) Explain the difference between average speed and instantaneous speed.

_____ [2 marks]

(b) A speed camera takes two photographs at half-second intervals of a car driving along a road. Explain how these could be used to calculate the speed of a driver.

_____
_____
_____ [3 marks]

Speed is not everything

1 (a) Explain what is meant by velocity.

_____
_____ [2 marks)

(b) Explain why the velocity of a car changes as it drives round a corner at a constant speed.

_____
_____ [2 marks]

(c) The car initially travels at 40 m/s. After braking, the car slows down and stops 5 seconds later. What is its acceleration?

_____
_____ [3 marks]

2 The velocity–time graph shows the motion of Alice on her bicycle.

(a) Describe Alice's motion in each section.

_____
_____ [4 marks]

(b) Explain how you could use the graph to calculate the distance Alice travels.

_____ [2 marks]

Forcing it

1 A van is trying to pull another car out of a ditch. Both vehicles cannot move.

(a) Describe how the force from the van compares with the force from the car.

_____

_____ [2 marks]

D–C

(b) A tractor is used to pull the car out of the ditch.
How does the force from the tractor compare with the force from the van?

_____ [1 mark]

2 (a) A car manual includes these tips for drivers to help them drive more economically:

- *do not drive at high speeds with your windows open – air conditioning is more efficient*
- *remove the luggage rack from the roof of your vehicle if you do not need it*
- *try to drive at a steady speed and avoid braking and accelerating.*

Use your ideas about forces to explain why these tips help to reduce fuel consumption.

_____

_____

_____

_____ [5 marks]

B–A*

Forces and acceleration

1 The table shows results from an experiment to investigate the link between force and acceleration.

(a) Describe the pattern shown in the results.

_____

_____ [2 marks]

Force	Acceleration
1 N	2 m/s^2
2 N	4 m/s^2
3 N	6 m/s^2
4 N	7 m/s^2
5 N	10 m/s^2

D–C

(b) Explain whether the results have been taken over a large enough range.

_____ [2 marks]

(c) Why should results from an experiment be repeated?

_____ [2 marks]

2 The mass of a rock is 4 kg.

(a) Calculate the force needed to accelerate the rock by 5 m/s^2.

_____

_____ [3 marks]

B–A*

(b) Explain how the mass of another rock could be found using a forcemeter and an accelerometer.

_____

_____ [3 marks]

Balanced forces

1 The diagram shows a tug of war between two teams. Neither team is winning so nobody is moving.

(a) Write down two pairs of balanced forces that act on the person labelled X.

(i) Pair 1 _____ and _____

(ii) Pair 2 _____ and _____ [4 marks]

X

(b) Describe how forces change when the team on the right hand side starts to win by pulling the other team towards them.

_____

[2 marks]

2 A person enters a lift and travels to the floor above. Describe and explain how the forces acting on this person change during this journey.

_____

_____

_____ [5 marks]

Stop!

1 (a) Stopping distance can be split into two parts. What are these called?

_____ [2 marks]

(b) Nicole is taking her first driving lesson. Why will her thinking distance be longer than her driving instructor's?

_____ [1 mark]

(c) The table shows how thinking distance changes with speed

Thinking distance (m)	0	6	9	12
Speed (km/h)	0	32	48	64

(i) Describe how thinking distance changes with speed.

_____ [2 marks]

(ii) A driver drinks a glass of wine and sometime later their thinking time is calculated as 1.0 seconds. Calculate their thinking distance when travelling at 9 m/s.

[1 mark]

(iii) A speed of 9 m/s is the same as 32 km/h. Use information from the table to describe the effect of the drink on the thinking distance of a driver.

_____ [1 mark]

2 Explain how the condition of a car can affect its stopping distance, giving examples. Explain which factor you think is most important.

_____

_____

_____ [6 marks]

Terminal velocity

1 (a) What is meant by terminal velocity?

_____ [1 mark]

(b) A ball bearing is dropped into a cylinder of oil.
Describe how each of the forces (including the resultant force) changes:

(i) when the ball bearing is first released in the oil.

_____ [2 marks]

(ii) when the ball bearing reaches terminal velocity.

_____ [2 marks]

D–C

2 (a) Explain in terms of forces why a skydiver slows down when they open their parachute.

_____
_____
_____ [5 marks]

(b) A designer claims to have invented a parachute that allows a skydiver to hover in calm weather conditions. Explain if this is possible in terms of forces.

_____
_____
_____ [4 marks]

B–A*

Forces and elasticity

1 Alex set up the experiment shown in the diagram to test the hypothesis that the extension of a spring depends on the force applied to it. The extension of the spring is its change in length.

(a) What is the independent variable _____ [1 mark]
(b) What is the dependent variable _____ [1 mark]
(c) Explain why it is important that Alex writes down all the data he collects, and not just the result of calculations.

_____
[1 mark]

(d) Explain two steps Alex should take to make sure his results are valid.

_____
[2 marks]

spring

mass

pointer

ruler

clamp and stand

D–C

2 A catapult manufacturer is testing different types of rubber to find the most suitable for their product. State two qualities of the rubber that would make it suitable for a catapult. Describe a test that the manufacturer could use to evaluate the most suitable choice.

_____
_____
_____ [5 marks]

B–A*

Energy to move

1 Complete these sentences:

(a) A moving car has kinetic energy transferred from _____ energy in its fuel. This energy is transferred to the surroundings in the form of _____ energy and _____ energy.

[3 marks]

(b) Explain why the energy transferred to the surroundings every second is less for a vehicle travelling on a smooth road than if it is travelling on a rough road at the same speed.

_____

_____ [2 marks]

D–C

2 A flywheel is a heavy wheel that stores kinetic energy when it spins. Evaluate factors that affect whether a flywheel could be used in a wind turbine to store energy when the wind is not blowing.

_____

_____

_____

_____ [5 marks]

B–A*

Working hard

1 Muhammad did 12 J of work when he lifted an apple.

(a) How much energy was transferred to the apple?

_____ [1 mark]

(b) What is the name of the force he was working against when he lifted the apple up?

_____ [1 mark]

D–C

(c) Muhammad dragged a box of apples 2 m along the ground. He measured the force needed as 25 N. How much work did he do?

_____

_____

_____ [3 marks]

2 A car is travelling at 30 m/s along a level road. At this speed, the frictional force is 600 m

(a) How far does the car travel in 5 s?

_____ [1 mark]

(b) How much work is done by the engine in this time?

_____

_____ [2 marks]

B–A*

(c) Explain why the engine must work harder when the car starts to drive up a slope at the same speed.

_____

_____ [2 marks]

Energy in quantity

1 A box is stored on a shelf in a warehouse.

  (a) Calculate the gravitational potential energy stored by a 6 kg box placed on a shelf 0.8 m above the ground.

  _____

  _____

  _____ [3 marks]

  (b) How much energy does the box have when it is lifted to a shelf that is 0.6 m higher?

  _____

  _____

  _____ [3 marks]

2 Explain which of these two objects has most kinetic energy: a person (mass 80 kg) jogging at 6 m/s or child on a skateboard (total mass 40 kg) moving at 8 m/s.

  _____

  _____

  _____

  _____

  _____

  _____ [6 marks]

Energy, work and power

1 A man uses a motor to lift some machinery. The motor lifts 400 kg of machinery by 3 m in 60 seconds.

  (a) What is the power of the motor?

  _____

  _____

  _____

  _____ [4 marks]

  (b) Explain what could change if the man uses a more powerful motor.

  _____

  _____

  _____

  _____ [4 marks]

2 Use your idea about work to explain why it is easier to pull a piano up a ramp compared with lifting it directly to the same final height.

  _____

  _____

  _____

  _____

  _____ [6 marks]

Momentum

1 The diagram shows two trolleys travelling towards each other and colliding. The two trolleys stick together after the collision.

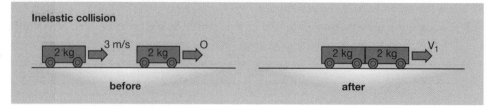

Inelastic collision

before

after

D–C

(a) Calculate the total momentum before the collision.

_____

_____ [3 marks]

(b) Write down the momentum after the collision.

_____ [1 mark]

(c) Calculate the speed that the joined up trolleys moves off with.

_____

_____ [3 marks]

B–A*

2 Explain how conservation of momentum allows the position of a satellite in space to be altered by small amounts using jet packs attached to the satellite.

_____

_____

_____

_____ [5 marks]

Static electricity

1 Owen rubs a balloon on some cloth and the balloon becomes charged.

(a) Explain how rubbing the balloon makes it negatively charged.

_____

_____

_____ [3 marks]

D–C

(b) What charge does the cloth have after it is rubbed on the balloon?

_____ [1 mark]

(c) How can the balloon be used to find the charge of another object?

_____

_____ [3 marks]

B–A*

2 (a) What is a gold leaf electroscope used to detect?

_____ [1 mark]

(b) Describe how a gold leaf electroscope can be used to show if a plastic comb has a positive or negative charge.

_____

_____ [3 marks]

Moving charges

1 A negatively charged balloon can stick on an uncharged painted wall.

  (a) What is the name of this effect?

  _____ [1 mark]

  (b) Explain why the balloon can stick on an uncharged painted wall.

  _____

  _____ [2 marks]

D–C

---

2 (a) Explain how electrical charge behaves differently in electrical conductors and in electrical insulators.

  _____

  _____ [2 marks]

  (b) An electrical sub-station contains equipment at very high voltages. A person inside the substation could receive an electric shock large enough to kill them without touching the equipment. Explain why.

  _____

  _____

  _____ [3 marks]

B–A*

Circuit diagrams

1 The diagram shows a series circuit and a parallel circuit, which use identical bulbs and cells. The ammeter reading in the series circuit is 0.4 A when the switch is closed.

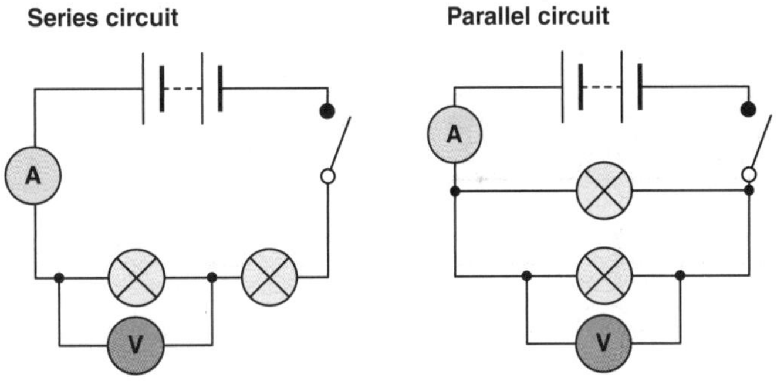

  (a) What is the ammeter reading in the parallel circuit when the switch is closed?

  _____ [1 mark]

  (b) The two bulbs are identical in the circuit, and each cell supplies 1.5 V.
  What is the voltage supplied to each circuit?

  _____ [1 mark]

  (c) Suggest a voltage reading for the voltmeter in the series circuit, giving a reason.

  _____

  _____ [2 marks]

Series circuit   Parallel circuit

D–C

---

2 Describe the energy transfers taking place in the series circuit.

  _____

  _____

  _____

  _____ [4 marks]

B–A*

Ohm's law

1 Alex measured the resistance of a piece of wire.

   (a) Explain what is meant by resistance.

   _____ [1 mark]

   (b) The resistance of the piece of wire was 10 ohms.
       Suggest a value for the resistance of a longer piece of the same wire.

   _____ [1 mark]

   (c) Write down two other factors that affect the resistance of a piece of wire.

   _____

   _____ [2 marks]

---

2 The graph shows the results of an experiment to find out how the current changes when the voltage changes in a wire.

   (a) What pattern does the graph show?

   _____

                             [1 mark]

   (b) Use the graph to write down the current when the voltage is 6 V.

   _____

                             [1 mark]

   (c) Use your answer to part b to calculate the resistance of the wire at 6 V.

   _____

   _____ [3 marks]

Non-ohmic devices

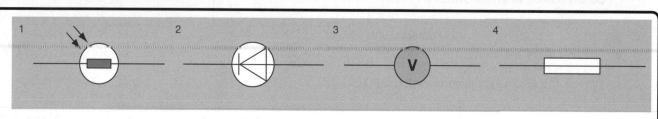

1 (a) How could you change the resistance of component 1?

   _____ [1 mark]

   (b) Write down one use for a circuit that contains component 1.

   _____ [1 mark]

   (c) A diode only allows the current to flow in one direction. Which symbol shows a diode?

   _____ [1 mark]

---

2 Semiconductor materials are used in different components. Explain two ways that circuits including these components respond to their surroundings.

   _____

   _____ [4 marks]

Components in series

1 The circuit shows three bulbs wired in series.

(a) Fred measures the current in four places in the circuit. What can you say about the readings?

_____ [1 mark]

(b) There are two cells each supplying 1.5 V. What is the value of the potential difference supplied to the circuit? ____V   [1 mark]

(c) The bulbs are all identical. Fred measures the potential difference across two of the bulbs. What is the reading on the voltmeter? _____ V   [2 marks]

(d) An extra bulb, identical to the others, is added to the circuit in series. Explain how the resistance of the circuit changes in as much detail as possible.

_____ [2 marks]

D–C

2 (a) A torch contained two batteries. One was inserted upside down. Explain how this affected the potential difference supplied to the circuit.

_____ [2 marks]

B–A*

Components in parallel

1 The circuit shows three bulbs wired in parallel. The circuit is switched on.

(a) If the current through each bulb is 0.5 A, calculate the current through the battery.

_____ [1 mark]

(b) The voltage supplied by the battery is 6 V. Write down the voltage across each of the three bulbs.

_____ [1 mark]

(c) Write down two advantages of wiring bulbs in parallel rather than in series.

_____ [2 marks]

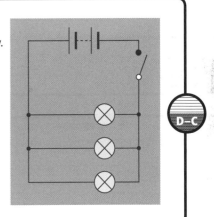

D–C

2 The diagram shows a circuit that includes three identical bulbs, which each have a resistance of 2 ohms. The battery supplies 6 V.

(a) Calculate the resistance of each path:

(i) upper path _____ ohms

(ii) lower path_____ ohms

(b) Calculate the combined resistance of both branches.

_____ [2 marks]

(c) Explain which branch will carry the largest current.

_____ [2 marks]

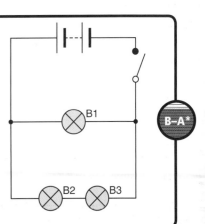

B–A*

Household electricity

1 The diagram shows two traces from an oscilloscope.

   (a) Which diagram shows the trace you would get from a battery?

   _____

   [1 mark]

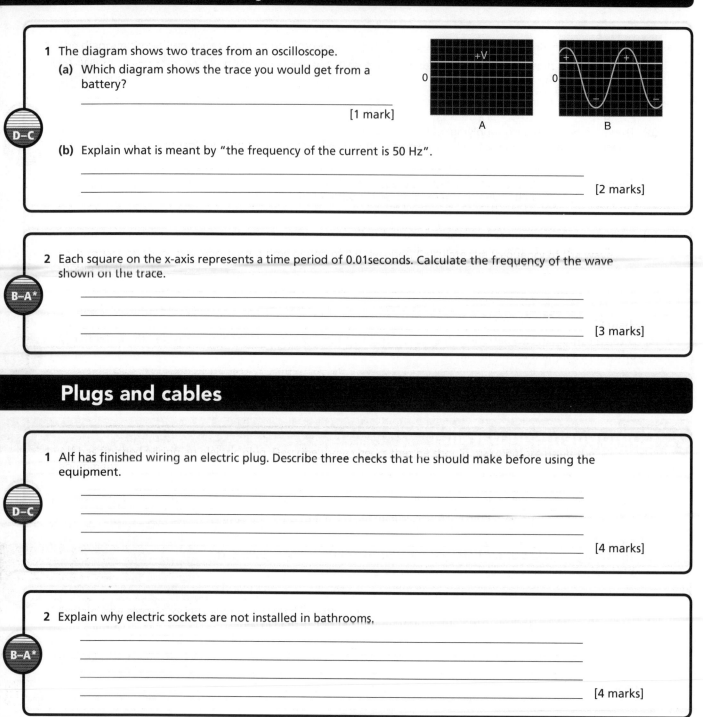

A   B

   (b) Explain what is meant by "the frequency of the current is 50 Hz".

   _____

   _____ [2 marks]

2 Each square on the x-axis represents a time period of 0.01seconds. Calculate the frequency of the wave shown on the trace.

   _____

   _____

   _____ [3 marks]

Plugs and cables

1 Alf has finished wiring an electric plug. Describe three checks that he should make before using the equipment.

   _____

   _____

   _____

   _____ [4 marks]

2 Explain why electric sockets are not installed in bathrooms.

   _____

   _____

   _____

   _____ [4 marks]

Electrical safety

1 When Alex was mowing the lawn using an electric lawn mower, he accidentally mowed over the cable, cutting it.

(a) A fuse box is installed in the house. Explain why Alex must not pick up the cable.

_____

_____ [3 marks]

(b) It is safer to use a residual current circuit breaker (RCCB) than to rely on a fuse box. Explain why.

_____

_____ [3 marks]

D–C

2 The power of an iron is 800 W. Nathan has a choice of fuses rated at 3 A, 5 A, 13 A.

(a) One fuse has 13 A printed on it. What does that tell you about the fuse.

_____

_____ [2 marks]

(b) Which fuse should Nathan use in the plug? Explain your choice.

_____

_____

_____ [3 marks]

(c) Why are fuses used?

_____

_____ [2 marks]

B–A*

Current, charge and power

1 The power of an electric fan is 1000 W.

(a) How much energy is transferred by the fan in 5 minutes?

_____

_____

_____ [3 marks]

(b) The current through the fan is 4.3 A. Calculate the charge transferred by the fan in 5 minutes.

_____

_____

_____ [4 marks]

D–C

2 Explain what affects the amount of energy transferred by electrical charge in a certain time.

_____

_____

_____

_____ [4 marks]

D–A*

Structure of atoms

1 (a) Complete the table to show the relative charges and masses of particles in the nucleus.

	Mass	Charge
electron	(i)	−1
proton	1	(ii)
(iii)	1	0

[3 marks]

(b) Carbon has several isotopes.

(i) What is different for different isotopes of carbon?

_____ [1 mark]

(ii) What is the same for all atoms of carbon?

_____
_____ [2 marks]

2 Describe how forces interact inside the atom.

_____
_____
_____
_____ [4 marks]

B–A*

Radioactivity

1 (a) What is meant by the term "radioactive"?

_____ [1 mark]

(b) Which form of radiation is most ionising: alpha, beta or gamma?

_____ [1 mark]

D–C

(c) Describe the changes that take place in the nucleus when a beta particle is emitted.

_____ [1 mark]

2 (a) Americium-241 decays by losing an alpha particle. Complete the values of X and Y in this equation.

$$_X^{241}\text{Am} \rightarrow \,_{93}^{Y}\text{Np} + \,_2^4\alpha$$

_____
_____ [2 marks]

B–A*

(b) Explain why smoke detectors containing Americium-241 are not a danger to people.

_____
_____ [2 marks]

More about nuclear radiation

1 The diagram shows one of the first models of the atom.

(a) Write down the three main features of the nuclear model of the atom.

_____

_____

_____

[3 marks]

positive 'jelly'

embedded electrons

(b) What evidence is there, that inside the atom:

(i) there is a positively charged nucleus,

_____ [1 mark]

(ii) the nucleus is very small,

_____ [1 mark]

(iii) electrons orbit outside the nucleus.

_____ [1 mark]

D–C

2 Explain why alpha, beta and gamma radiation behave differently in electromagnetic fields.

_____

_____

_____

_____ [4 marks]

B–A*

Background radiation

1 The diagram shows the sources of background radiation in the UK.

(a) What proportion of background radiation comes from radon and thoron?

_____

[1 mark]

(b) Explain why people may be exposed to different levels of background radiation.

_____

_____

_____

[3 marks]

0.5% fallout from nuclear weapons testing

0.1% total discharge from nuclear power industry

11.5% medical, mainly X-rays

0.4% work related

0.5% other - mainly air travel

14% cosmic rays from outer space

17% radiation from the bodies of all living things and their food

19% gamma rays from rocks and soil

radon and thoron gas released from soil, rocks and building materials

D–C

2 Explain whether background radiation is a significant health risk.

_____

_____

_____

_____ [5 marks]

B–A*

Half-life

1 (a) Explain what is meant by half-life.

_____

_____

_____ [3 marks]

(b) The proportion of radioactive carbon in a wooden arrow has been measured. Explain whether the proportion of radioactive carbon in an arrow 10 000 years old will be more or less than that found in an arrow made from modern wood.

_____

_____ [2 marks]

2 A sample of cobalt has a half-life of 5 years. Its count rate is measured as 1200 counts per second.

(a) 5 years later, what will the count rate be?

_____ [1 mark]

(b) How long will it take for the count rate to fall from 1200 counts per second to 150 counts per second?

_____

_____

_____ [3 marks]

Using nuclear radiation

1 The diagram shows one method of controlling thickness of cardboard in a factory.

(a) Write down one reason why it is important to control the thickness of cardboard.

_____ [1 mark]

(b) Why can't alpha radiation be used to monitor the thickness of cardboard?

_____

_____ [1 mark]

(c) Explain what adjustments the machine should make to the separation of the rollers if the amount of radiation reaching the detector decreases.

_____ [2 marks]

2 Explain whether there are any ethical issues with the medical use of radioactive tracers.

_____

_____

_____ [5 marks]

Nuclear fission

1 (a) What is meant by nuclear fission?

_____

_____ [2 marks]

(b) Write down one use for nuclear fission in the UK.

_____ [1 mark]

(c) The diagram shows a chain reaction.
How many neutrons are produced from the first stage in this chain reaction?

_____ [1 mark]

(d) Explain why a chain reaction must be controlled.

_____

_____

_____ [3 marks]

D–C

---

2 Pace makers are inserted in a patient with irregular heartbeats. They control the heartbeat so it beats in a regular pattern. Suggest advantages and disadvantages of using a small nuclear power source for pace makers.

_____

_____

_____

_____ [4 marks]

B–A*

Nuclear fusion

1 (a) Describe two differences between nuclear fission and nuclear fusion.

_____

_____ [2 marks]

(b) Write down one place in our solar system where nuclear fission takes place.

_____ [1 mark]

D–C

---

2 Explain how nuclear fusion can explain the presence of different elements in the Universe.

_____

_____

_____

_____

_____ [5 marks]

B–A*

Life cycle of stars

D–C

1 Describe how the forces acting in a star control how it moves between different stages in its life cycle.

_____

_____

_____

_____ [5 marks]

B–A*

2 Explain why the size of a star affects its life cycle.

_____

_____ [3 marks]

Extended response question

Describe two safety features, found in modern cars, which are designed to protect the driver and passengers in the event of a crash. Use your knowledge of energy and momentum to explain how each feature protects people in the car from serious injury or death.

The quality of written communication will be assessed in your answer to this question.

_____

[6 marks]

Using X-rays

1 (a) The box contains some statements about X-rays. Tick the box for each statement that is correct. [4 marks]

Statement	Tick
X-rays cause ionisation	
X-rays are absorbed by bone but pass through skin	
X-rays have long wavelengths	
X-rays are absorbed by all tissues	
X-rays are used to kill cancer cells	
X-rays do not cause ionisation	
X-rays have a very high frequency	

(b) X-rays can be detected using photographic film and using charge-coupled devices linked to computers. Explain one advantage of using charge-coupled devices to look at X-ray images.

_____ [2 marks]

2 People working with X-rays need to take precautions. Describe two suitable precautions and explain how they reduce the risk of harm.

_____

_____ [4 marks]

Ultrasound

1 (a) Use these words to complete the sentences.

Words may be used once, more than once or not at all.

absorbed   analysed   reflected   shadow   transmitted

X-rays are _____ by bone and _____ through soft tissue.

This creates a _____ picture that can be used for a medical diagnosis.

Ultrasound is partly _____ and partly _____ at a boundary.

Reflections are _____ at a detector to create an image. [5 marks]

(b) Explain one advantage of using ultrasound for prenatal scanning.

_____

_____ [3 marks]

2 The diagram shows an ultrasound trace taken to measure the distance to the back of a patient's eyeball.

(a) Explain why there are two peaks. [2 marks]

(b) Explain why one peak is smaller than the other peak. [2 marks]

(c) Use the trace to calculate the distance to the back of the person's eyeball.

Each square along the x axis represents 0.02 m. Ultrasound travels at 1500 m/s in eye tissue.

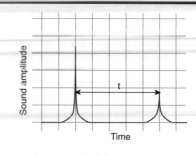

_____

_____ [4 marks]

Refraction

1 A lens has a focal length of 0.1 m.

Convex

focal point

D–C

(a) Calculate the power of the lens in dioptre.

_____
_____
_____
_____ [2 marks]

(b) Explain whether a more powerful lens has a longer or a shorter focal length.

_____
_____
_____
_____ [2 marks]

2 An optician is choosing materials to make spectacles for a patient. The patient needs a very powerful lens and one much less powerful lens.
Explain how the optician can make both lenses the same thickness.

_____
_____
_____
_____
_____
_____
_____
_____ [4 marks]

B–A*

Lenses

1 A magnifying glass is used to look at a picture of a tree. The picture of the tree is 1 cm high. The image of the tree viewed through the magnifying glass is 5.5 cm.

    (a) What is the magnification of the image?

_____

_____ [2 marks]

    (b) Will the tree appear smaller or taller if the magnifying glass is moved further away from the picture?

_____ [1 mark]

D–C

2 The diagram shows the ray diagram for a convex lens.

    (a) Describe the image in as much detail as you can.

_____

_____

_____ [3 marks]

    (b) Describe how the image changes if the object is moved closer to the lens than the focal point.

_____

_____

_____ [3 marks]

B–A*

Seeing clearly

1 The diagram shows the structure of an eye.

    (a) Name the part of the eye that detects light.

_____ [1 mark]

    (b) Explain how the shape of the lens in the eye can be changed.

_____

_____

_____ [3 marks]

D–C

2 Alex visited an optician. He could see nearby objects clearly, but distant objects were blurred.

    (a) Did Alex have long or short sight?

_____ [1 mark]

    (b) Explain how Alex's eyesight can be corrected using lenses.

_____

_____

_____ [3 marks]

U–A*

More uses of light

1 (a) Explain how light travels through an optic fibre.

_____

_____ [2 marks]

(b) Complete the diagram to show the path of light through the optic fibre.

glass

light ray

[3 marks]

D–C

---

2 Calculate the refractive index for glass. Its critical angle is 41°.

_____

_____

_____ [3 marks]

B–A*

Centre of mass

1 Explain how to find the centre of mass of a tennis racket.

_____

_____

_____ [3 marks]

D–C

---

2 Explain how a lorry's trailer should be stacked to make sure it is less likely to topple.

_____

_____

_____

_____ [4 marks]

B–A*

Moments

1 Two people are pushing on a door. One person is pushing the door open using a force of 30 N, 0.7 m from the hinge.

(a) Calculate the moment of this force.

_____

_____ [2 marks]

(b) A second person pushes on the door in the opposite direction, 0.5 m from the hinge. The door is not moving. Write down the moment of this force.

_____ [1 mark]

(c) Calculate the force that the second person is using on the door.

_____

_____ [2 marks]

2 The diagram shows a crane. It has a counterweight and can lift loads on two hooks.

The counterweight weighs 60,000 N.

(a) Calculate its moment.

_____

_____ [2 marks]

(b) What is the heaviest load that can be lifted safely on the hook that is 30 m from the pivot?

_____

_____ [2 marks]

(c) Another load is lifted at the same time, using the second hook. Explain two changes that should be made to ensure the crane makes the lift safely.

_____

_____ [5 marks]

Hydraulics

1 The diagram shows a simple hydraulic system. A pair of syringes is linked using a tube filled with water.

A force of 25 N is applied to the left-hand syringe. The area of the left-hand piston is 0.001 m².

(a) Calculate the pressure inside the system.

_____

_____ [3 marks]

(b) The area of the right hand piston is 0.005 m². Calculate the force that the right hand piston supplies

_____

_____ [2 marks]

2 (a) Explain two benefits of using a hydraulic system in a braking system in a car.

_____

_____ [4 marks]

(b) Brakes need to work consistently at all temperatures. Explain two properties that a specialised brake fluid should have.

_____

_____ [4 marks]

Circular motion

1 The diagram shows a roundabout designed for
young children. A child sits on the seat at the
centre of the roundabout.

(a) What is the name of the force felt by a child
on the roundabout when it spins?

_____ [1 mark]

(b) Describe how the force felt by the child and the direction of movement changes when:

i) the roundabout spins faster.

_____

_____ [2 marks]

ii) the child moves to sit on one of the seats on the outer edge of the roundabout.

_____

_____ [2 marks]

2 A car accelerates as it drives at a steady speed around a bend. Explain why this statement is correct.

_____

_____ [2 marks]

Circular motion in action

1 A child on a swing swings 30 times in 1 minute.
(a) What is the time period of their swing?

_____ [1 mark]

(b) What is the frequency of their swing?

_____

_____ [3 marks]

(c) Explain whether the frequency changes if the child swings higher.

_____ [1 mark]

2 A fairground ride includes a giant swing. Describe the energy changes that take place when the ride is in motion.

_____

_____

_____ [3 marks]

Electromagnetic force

1 The diagram shows a simple electric motor.
Complete the sentences to explain how the motor works. Match words A–D with the numbers 1–4 in the sentences.

A = axle   B = coil
C = current   D = magnetic field

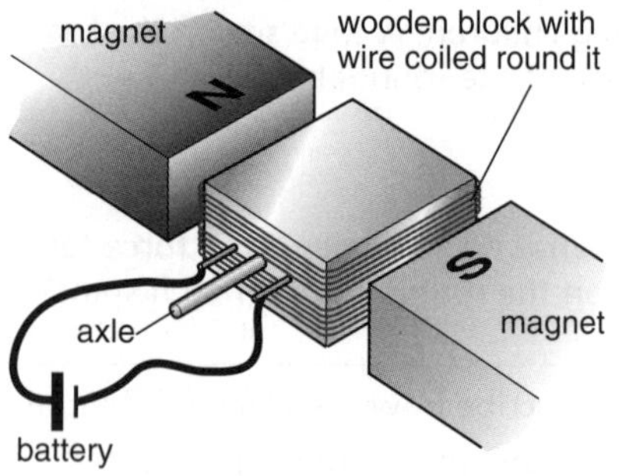

A _____1_____ flows around the coil of wire that can rotate.

The current flows at right angles to the _____2_____ .

The _____3_____ feels an upwards force on one side and a downwards force on the other side.

The coil of wire spins around the _____4_____ .

[4 marks]

(diagram labels: magnet, N, wooden block with wire coiled round it, S, axle, magnet, battery)

2 Explain how the motor effect is used in an electric drill.

_____

_____

_____

_____ [4 marks]

Electromagnetic induction

1 A student is carrying out an experiment. She is moving a magnet into a coil of wire.

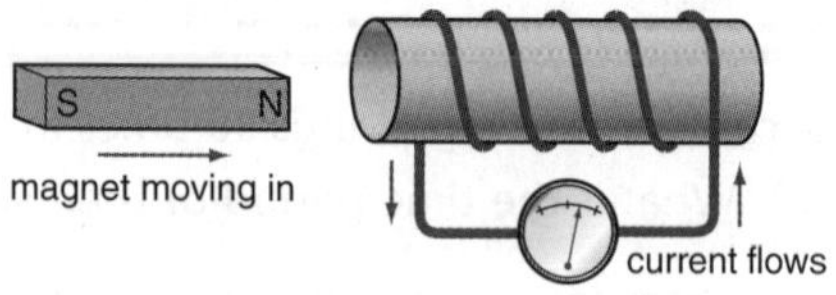

(a) An ammeter connected to the coil of wire registers a current. What happens to the ammeter reading when she stops moving the magnet?

_____ [1 mark]

(diagram labels: S N, magnet moving in, current flows)

(b) What happens to the ammeter reading when she pulls the magnet out of the coil of wire at the same speed?

_____

_____ [2 marks]

(c) Describe two changes she could make to her experiment to increase the ammeter reading.

_____

_____ [2 marks]

2 A fuse is designed to melt if the current through it is larger than a certain value. Fuses usually melt just as equipment is turned on. Use ideas about electromagnetic induction to explain why.

_____

_____

_____ [3 marks]

Transformers

1 The diagrams shows a transformer.

primary coil   secondary coil

(a) Explain whether this is a step-up or a step-down transformer.

_____

_____ [2 marks]

(b) Explain why the coils of wire must be made from insulated wire.

_____ [1 marks]

(c) Explain why the transformer only works with AC current.

_____

_____ [2 marks]

D–C

2 A transformer has 1000 turns on its primary coil, and 25 000 turns on the secondary coil. Calculate the output voltage if the input voltage is 12 V.

_____

_____

_____

_____ [4 marks]

B–A*

Using transformers

1 A transformer used in a factory is assumed to be 100% efficient.

(a) Explain what 100% efficient means.

_____ [1 mark]

(b) The current flowing in the transformer is 1.6 A when the input voltage is 400 V.
Calculate the power input to the transformer.

_____

_____

_____ [3 marks]

D–C

(c) Write down the power output of the transformer.

_____ [1 mark]

(d) If the voltage output of the transformer is 120 V, calculate the current in the secondary coil.

_____

_____ [2 marks]

2 Switch-mode transformers have several advantages. Explain whether switch-mode transformers should be installed in existing equipment, replacing existing transformers.

_____

_____

_____

_____

_____ [5 marks]

B–A*

Extended response question

X-rays and ultrasound scans are used to diagnose and treat different medical conditions.

Explain why the different types of scans are useful in different circumstances.

The quality of written communication will be assessed in your answer to this question. ✎

_____

_____

_____

_____

_____

_____

_____

_____

_____

_____

_____

_____

_____

_____

_____

_____

_____

_____

_____

_____

_____

_____

_____

_____

_____ [6 marks]

P1 Physics Checklist

1. I can compare how heat is transferred by conduction, convection and radiation ☐
2. I know what factors have an effect on the way we heat and insulate buildings ☐
3. I can use kinetic theory to explain the properties of different states of matter ☐
4. I know what is meant by efficient use of energy, and how to compare the efficiency of energy transfers ☐
5. I can calculate the energy transferred by electrical appliances ☐
6. I can describe different methods of generating electricity including renewable and non-renewable energy sources ☐
7. I can evaluate the most suitable ways of generating electricity in different circumstances ☐
8. I can explain why electricity is transmitted at high voltages in the National Grid ☐
9. I can describe properties of transverse waves, including reflection and refraction ☐
10. I can describe the uses and hazards of electromagnetic waves ☐
11. I can calculate the speed of waves ☐
12. I can describe the properties of longitudinal waves such as sound waves ☐
13. I know what is meant by red-shift ☐
14. I can describe the Big Bang theory and explain why scientists believe the Universe is still expanding ☐

P2 Physics Checklist

1. I can calculate a resultant force, and describe its effect on objects including elastic objects ☐
2. I can calculate the acceleration of an object ☐
3. I can draw and interpret distance time graphs and velocity time graphs ☐
4. I can describe different factors that affect the stopping distance of a vehicle ☐
5. I can explain why a moving object reaches a terminal velocity ☐
6. I can carry out calculations involving power and energy ☐
7. I can explain momentum changes during collisions and explosions using calculations ☐
8. I can describe how a static electric charge builds up and its effects ☐
9. I know standard circuit symbols and can draw and interpret voltage-current graphs for different components ☐
10. I can use Ohm's law ☐
11. I can calculate current, voltage and resistance in series and parallel circuits ☐
12. I can describe safe practices in the use of electricity ☐
13. I can describe the structure of atoms ☐
14. I can compare different properties of ionising radiation ☐
15. I know the origins of background radioactivity ☐
16. I can carry out half-life calculations ☐
17. I know what nuclear fission is and what is meant by a chain reaction ☐
18. I know what nuclear fusion is and how it is involved in the formation of different elements ☐
19. I can describe the life cycle of stars ☐

1. I can compare the medical use of ultrasound and X-rays ☐

2. I can use an oscilloscope trace to calculate the distance to a boundary ☐

3. I can describe the advantages and disadvantages of using ultrasound, X-rays and computerised tomography (CT) scans ☐

4. I can draw ray diagrams and describe the image formed, including its magnification ☐

5. I can calculate the refractive index of a material ☐

6. I can choose the most suitable lenses to correct defects of vision ☐

7. I can calculate the power of a lens ☐

8. I can describe the structure of the eye ☐

9. I can describe uses of a laser and of optical fibres ☐

10. I can find the centre of mass of different objects ☐

11. I know what affects the stability of an object and can predict if it is likely to topple ☐

12. I can calculate the moment of a force ☐

13. I can describe how a hydraulic system works as a force multiplier ☐

14. I can interpret and evaluate data on objects moving in circular paths ☐

15. I know what is meant by the motor effect and what affects it ☐

16. I can explain how electromagnetic appliances work using diagrams ☐

17. I can compare the effectiveness of different types of transformer for a particular use ☐

18. I can describe the advantages of switch mode transformers ☐

Answers

P1 Answers

Pages 66–67

Energy

1a Electrical energy [1]; changes to heat [1]; kinetic energy [1]; and sound [1].

b It is transferred to the surroundings. [1]

c Olympus 1000 [1], it took least time to dry the cloth [1].

d Any two valid points, for example: distance from hairdryer to cloth [1]; dampness of cloth initially [1]; dryness of cloth finally [1]; size of cloth [1].

2 More energy transfers are involved [1], as electricity is used to recharge the batteries [1]. Energy is wasted at each transfer [1], so more energy is wasted [1].

Infrared radiation

1a Black [1]. Solar panels absorb infrared radiation from the Sun [1]. Black is the best colour to absorb infrared radiation [1].

b Any two from: radiation from the sun is more intense from the south [1], more infrared radiation will be absorbed [1], the water will heat up quicker [1].

c The house loses heat by radiation when it is warmer than the surroundings [1]. At night/during winter, it is cooler outside the house compared to inside [1]. Black surfaces emit infrared radiation well [1].

2a Yes [1], the beaker absorbs and emits infrared at the same rate [1].

b More infrared is emitted than absorbed [1] so the beaker cools down [1], until it reaches the same temperature as the fridge [1] when infrared is absorbed and emitted at the same rate [1].

Kinetic theory

1a Melting is when a solid changes to a liquid. [1]

b Particles are in fixed positions in a solid [1] but change places in a liquid [1]. In a solid, ice bonds hold the particles together [1]. When ice melts, bonds between particles start to break and reform as the ice melts to liquid [1].

c Energy is needed to break bonds [1]. Energy is absorbed from the surroundings [1].

d Energy is absorbed more quickly. [1]

2 Any four from: particles in solids are held by strong bonds [1]; but particles in air have absorbed enough energy to break bonds [1]; particles in solids vibrate slightly in fixed positions [1]; but particles in air move rapidly [1]; copper particles are heavier than particles in air (mainly nitrogen and oxygen) [1].

Conduction and convection

1a Convection is when particles move through a substance, transferring energy [1]; particles in solids cannot change places and move through a solid [1].

b The flask reduces conduction [1] with the poly(ethene) base/vacuum/insulated stopper, which is an insulator [1]. Reduces convection [1]: through the top using the stopper [1] OR through the sides using the vacuum [1].Reduces radiation [1] with the silvered walls [1].

2 Heat sinks need to conduct heat rapidly away from components [1]. Metals contain free electrons [1] and conduct heat faster than plastics [1].

Pages 68–69

Evaporation and condensation

1a Evaporation is when a liquid changes to a gas. [1]

b Boiling takes place throughout a liquid; evaporation takes place at the surface. [1] Boiling takes place at the boiling temperature, evaporation takes place at any temperature. [1]

c Any two from: warmer weather – molecules have more energy and can break free of bonds more easily [2], windy or drier weather – surrounding air does not become saturated [2].

2 Credit sensible methods.

Equipment: cotton wool, thermometer, stopwatch [1]; water, aftershave [1].

Method: soak cotton wool pads in water and aftershave [1]; monitor temperature of each over a period of time [1].

Control/change variables: Use same sized pads and soak up same mass of liquid [1]; keep pads under same conditions [1].

Rate of energy transfer

1 If the potato is cut, the surface area is larger [1] and the distance to the centre is smaller [1]. More heat can be transferred by radiation at the surface [1]. Heat is not transferred as far by conduction through the potato [1].

2 Heat is transferred quicker if there is a larger temperature difference [1]. Boiling oil is hotter than boiling water [1].

Insulating buildings

1a The report can suggest savings that are greater than the cost of the report. [1]

b Total savings are £275 per year [1] or £1375 in 5 years [1]. Overall savings are £1080 [1].

c Loft insulation has a low U-value [1] as it is a good insulator/reduces heat losses [1].

d Maximum of 2 marks for two suggestions, and maximum 2 marks for two explanations, for example: cavity wall insulation or double glazing [1] reduces heat losses by conduction [1]; installing draught excluders [1] reduces losses by convection [1]; putting foil behind radiators [1] reduces heat losses by radiation [1].

2 Credit any valid response with explanations, for example:

Yes [1], the council has limited money [1]; it is better to have some energy-saving measures in as many homes as possible [1]; installing double glazing may use more energy in production and installation [1] than it saves [1].

No [1], badly fitting windows waste a large amount of energy [1]; in some cases the savings will be greater than the average shown in the table [1]; double glazing saves some energy every year [1]; it will last for a long time [1].

Specific heat capacity

1a Specific heat capacity is the energy absorbed by 1 kg of a material when its temperature increases by 1 °C. [1]

b Temperature rise = energy/(mass x specific heat capacity) [1] = 1800/(0.25 x 900) [1] = 8 °C [1]

2 A saucepan needs to heat up quickly and evenly [1]. Copper has a lower specific heat capacity so it heats up quicker than aluminium [1], so copper would be a better choice [1].

Pages 70–71

Energy transfer and waste

1a chemical energy -> kinetic energy + heat energy (2 marks if all forms of energy are correct; 1 mark if any form of energy is incorrect or missing)

b As wasted heat energy. [1]

c Conservation of energy states that energy is not lost or created during an energy transfer. [1] The width of the arrows shows the relative proportion of each form of energy [1]. The total width of the output arrows equals the width of the input arrow [1].

2

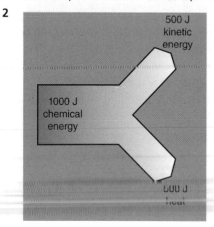

Give marks as follows: a double headed arrow [1]; equal width arrows [1]; correctly labelled energy forms [1]; correctly labelled amounts of energy [1].

Efficiency

1a efficiency = useful output energy/input forms of energy [1] = 60% [1]

b Allow sensible answers, for example: Yes [1], more efficient equipment saves money on running costs [1] so people should have this information [1].

OR No [1], Sankey diagrams are hard to interpret/contain too much information [1]; the efficiency could be stated as a percentage (%) [1].

2 No energy transfer can be more than 100% efficient [1]. There is more than one energy transfer if the motor is battery operated [1]. Energy is wasted at each transfer [1].

Electrical appliances

1a chemical (from food [1] -> kinetic (winding the recharger) [1] -> chemical (in battery) [1]

b i For example: The recharger does not use mains electricity [1], so it can be used where there is no mains supply [1].

 ii For example: A battery is a portable supply of electrical energy [1], so the phone can be used on the move [1].

2a Credit valid answers, any two from: the expense of providing a mains supply of electricity to the garden [1]; safety issues if the supply is wrongly installed [1]; it is inconvenient to provide a mains supply [1]; there are portable alternatives [1].

b Any two from, for example: solar panels need no maintenance [1]; there is sufficient sunlight outside to power lights [1]; batteries are expensive to replace [1].

Energy and appliances

1 energy = power x time [1] = 850 W
850 x 15 x 60 seconds [1] = 765 000 J [1]

2 The microwave oven is less powerful than the electric oven [1] and uses a smaller current [1], so the cable doesn't heat up as much [1] so a thinner cable is sufficient [1].

Pages 72–73

The cost of electricity

1a 0.1 x 1500 [1] = 150 kWh [1]

b 150 kWh at 15p [1] = £22.50 or 2250p [1] x 5 years = £112.50 [1]

c For example: less electricity is used lighting homes [1]; generating less electricity reduces carbon emissions from power stations [1]

2 Credit valid responses, for example: This saves the householder money [1] as they are not paying for electricity used to keep equipment on standby [1]. Less electricity will be wasted so less electricity needs to be generated [1], which reduces carbon emissions [1]. However, there is a cost in buying new equipment [1]. Manufacturing new equipment uses electricity/causes carbon emissions, etc. [1]. (maximum 6 marks)

Power stations

1a chemical in gas -> thermal in water and kinetic in steam -> kinetic in turbine and generator -> electrical in generator (1 mark per correct transfer)

b Unwanted heat from combined heat and power stations is used to directly heat homes [1]. In power stations which only generate electricity, electricity is transferred to homes then used for heating [1] there are more stages in the energy transfer [1] energy is lost at each stage. [1]

2 One mark for stating an option and three marks for valid reasons. Credit valid responses, for example: Convert existing older power stations [1]: fewer resources will be needed [1]; quicker timescale [1]; lower overall cost [1].

or

Develop new technologies: [1] older power stations will never match the efficiencies of new ones [1]; existing fuels will run out in future [1]; so we need new/renewable/non-polluting sources of energy [1].

Renewable energy

1 Iceland uses geothermal energy [1]. This uses heat in volcanic rocks to produce steam to spin turbines [1]. Norway uses hydroelectricity [1]. This uses falling water trapped behind dams to spin turbines [1].

2 Credit valid responses, for example: Advantages include being able to use free, renewable energy sources; using batteries increases our use of wind power which reduces our greenhouse gas emissions [1]. Disadvantages include the expense of installing batteries; difficulties of transferring energy from batteries to homes [1].

Electricity and the environment

1

Type of station	Advantage of the scheme	Disadvantage of the scheme
Coal-fired	a single power station in a relatively small land area generates large amounts of energy [1]	greenhouse gas emissions; mining damages the environment; large quantities of waste to dispose of [1]
Wind turbines	no pollution is emitted to the atmosphere [1]	wind farms take up large land areas; noise pollution [1]

2 Two reasons from: Harm caused to wildlife [1] as local habitats are destroyed to grow crops [1]; reduced biodiversity [1] caused by crop monoculture [1]; burning biofuels [1] releases greenhouse gases [1]. (maximum 4 marks)

Pages 74–75

Making comparisons

1a Credit sensible reason and explanation, for example: more demand for lighting at night because it is dark [1]; more demand for electricity for heating in winter because it is cold [1]; more demand for electricity at meal times for cooking [1]. (3 marks maximum)

b Waves, wind, tidal, gas. [3]

2 Credit sensible answers that are linked to the information. These are examples of possible responses: Wind farms are unreliable [1] and fossil fuel power stations are reliable [1]. It's more important to have a reliable supply of electricity [1] so we should only have a few wind farms [1] in carefully chosen places which are really windy [1]. OR It's more important to reduce reliance on fossil fuels [1] so we should have a large numbers of wind farms [1] but improve their reliability [1]. (maximum 5 marks)

The National Grid

1a Smaller current [1] so less energy is wasted as heat. [1]

b current = power/voltage = 200 000 000 / 400 000 [1] = 500 [1] A [1].

2 For example: Bury underground [1]: the extra cost is worth paying to reduce environmental impact [1]. It reduces the risk of damage during severe weather [1] and harm due to accidents [1].

OR Keep as overhead power lines [1]: it is too expensive/disruptive to change [1], the landscape will still be damaged with underground lines [1], easier to find and mend faults with overhead power lines [1]. (maximum 4 marks)

What are waves?

1 Credit valid answers

Difference 1: vibrations are perpendicular to direction of energy travel (transverse) or parallel to direction of energy travel (longitudinal). [1]

Difference 2: longitudinal waves cannot transfer energy through a vacuum but transverse waves can. [1]

Similarity 1: both waves transfer energy. [1]

Similarity 2: both types of wave travel from a source. [1]

2a Mechanical waves cannot transmit energy through a vacuum. [1]

b Waves transfer energy from the earthquake to other places [1]. The time taken depends on the distance from the earthquake to other places [1]. Waves travel at different speeds in different materials [1]. Longitudinal waves travel at different speeds to transverse waves [1].

Changing direction

1 Diffraction is the spreading of waves through a gap [1]. Diffraction is most pronounced when the wavelength of the wave is similar to the gap [1]. Doorways are a similar size to the wavelength of many sound waves [1].

2 When light is refracted, it changes direction at a boundary [1]. Light changes direction in water compared with air [1]. To the observer, the light ray appears to have travelled in a straight line rather than changing direction [1].

Pages 76–77

Sound

1a Any number between 441 Hz and 480 Hz. [1]

b It gets shorter. [1]

c The amplitude [1] gets bigger [1].

2a The trace shows more waves [1] that have a smaller amplitude [1]. The waves should be the same height and have the same wavelength throughout.

b Credit valid responses, for example: oscilloscopes provide information about different sounds [1]. Louder sounds will have a larger amplitude and taller trace on the screen [1]. The engineer can use the information to adjust the position/settings on the microphone near that player [1].

Light and mirrors

1 The image is upright, virtual and the same distance behind the mirror. [3]

2

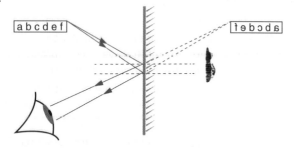

One mark for each ray linking the eye and the object [2]. (Credit rays correctly drawn from 2 different places on the object that link correctly to corresponding place on the image.) 1 mark for extending rays from mirror to image [1]. 1 mark for adding arrows to the rays [1].

Using waves

1

Type of electromagnetic radiation	Use
Radio wave	BBC TV broadcasts
Microwave	Satellite TV
Infrared radiation	Remote control
Visible light	Photography

(1 mark for each correct pair)

2 Mobile phones use microwaves [1]. Microwaves have shorter wavelengths than radio waves [1]. The wavelength of microwaves is much smaller than the size of hills and buildings [1]. Microwaves do not diffract round hills and buildings [1]. Phones must be in line-of-sight of transmitters for good reception [1].

The electromagnetic spectrum

1 wavelength = wavespeed/ frequency [1] = 300 000 000/ 2 700 000 000 [1] = 0.111 [1] m [1] (maximum of 3 marks if units are not correctly converted)

2 The electromagnetic spectrum is a continuous spectrum of waves whose wavelength varies from more than 10 000 m to 10^{-15} m. [2]

Pages 78–79

Dangers of radiation

1a Ultraviolet. [1]

b Skin cancer [1]; sunburn [1], premature aging on the skin [1].

c Wear a hat/cover up [1] so radiation cannot reach the skin [1]; stay inside between the hours of 11 am and 3 pm [1] to avoid very intense sunlight [1].

2a The DNA molecules are ionised. [1]

b Cancer cells are killed [1], healthy cells outside the treatment area receive a lower dose/no radiation [1]; healthy cells are less likely to be damaged [1].

Telecommunications

1 Microwaves transmit images to satellites [1] satellites transmit the images to receiving stations in Australia [1]. Microwaves travel so fast (300 million m/s) that the signal arrives in less than a second [1].

2 The satellite takes 24 hours for one complete orbit [1]. It orbits in the same direction as Earth rotates, so stays above the same point all the time [1]. The satellite dish can be fixed facing the direction of the satellite [1].

Cable and digital

1a Analogue. [1]

b The signal needs to be amplified/gets weaker when it is transmitted long distances [1]; when the signal is amplified, any interference is amplified too [1].

Searching space

1a Different objects emit different types of electromagnetic radiation [1]; the same detectors cannot detect all forms of electromagnetic radiation [1].

b It is larger [1] because radio waves are longer than visible waves [1]

Page 80

Waves and movement

1a Doppler effect. [1]

b The wave has a longer wavelength [1] but the same amplitude. [1]

2a The red shift is when light from an object moving away from an observer [1] appears to have a longer wavelength/appears redder [1].

b The larger the red shift, the faster the object is moving away [1]; more distant galaxies have larger red shifts than closer galaxies [1].

Origins of the Universe

1a All matter and energy in the Universe were in one place at one point in time [1]. About 14 billion years ago, a rapid expansion started to take place [1], which is still continuing [1].

b Cosmic microwave background radiation/the echo of the Big Bang [1]. The red shift which provides evidence that galaxies are all moving apart [1]. The larger red shift from more distant galaxies, which is evidence more distant galaxies are moving apart faster [1].

2 Cosmic microwave background radiation cannot be explained using other theories [1]. It is the same frequency we would expect radiation from the initial expansion to have if the theory is correct [1]. It is detected coming from all directions with equal intensity [1].

Page 81

Extended response question

5 or 6 marks:

A detailed description of relevant properties of radio waves, microwave, infrared and visible light (e.g. they all travel at the speed of light; some are absorbed by the atmosphere; diffraction effects are noticeable with radio waves). The communication use of these electromagnetic waves is linked to its properties, e.g. microwaves are not absorbed by the atmosphere so can be used for satellite communication. *All information in answer is relevant, clear, organised and presented in a structured and coherent format. Specialist terms are used appropriately. Few, if any, errors in grammar, punctuation and spelling.*

3 or 4 marks:

A limited description of relevant properties of radio waves, microwave, infrared and visible light. Some communication uses of members of the electromagnetic spectrum are stated, but may not always be linked to its properties. *For the most part the information is relevant and presented in a structured and coherent format. Specialist terms are used for the most part appropriately. There are occasional errors in grammar, punctuation and spelling.*

1 or 2 marks:

The answer includes an incomplete description of relevant properties, which may be linked to specific members of the electromagnetic spectrum. Some relevant uses of more than one type of electromagnetic wave should be stated. *Answer may be simplistic. There may be limited use of specialist terms. Errors of grammar, punctuation and spelling prevent communication of the science.*

P2 Answers

Pages 82–83

See how it moves

1a i travelling forward at a steady speed [1] ii stopped [1]

b section C [1]

c a curved line getting steeper [2]

2a Instantaneous speed is speed at a certain point in time (e.g. seen on a speedometer) [1]. Average speed is distance/time for the whole journey [1].

b The photographs are used to compare the position of the car relative to markings at known intervals [1]. The distance travelled by the car is calculated [1]. Use speed = distance/0.5 s to calculate speed [1].

Speed is not everything

1a Velocity is speed in a certain direction. [2]

b Since velocity depends on speed and direction [1], although speed remains constant, the car's direction changes [1].

c Acceleration is change in speed/time [1]

$$\frac{-40}{5} = -8 \; [1] \; m/s^2 \; [1]$$

2a Section A – steady speed [1]; section B – accelerates [1]; section C – decelerates [1] section D – slower steady speed [1]

b Work out the area under each section of the graph [1] and add the distances together [1].

Forcing it

1a The car and van pull in opposite directions [1] with equal forces [1].

b The tractor pulls harder / force is larger than the van [1].

2 Any five from: drag forces are larger if there is a roof rack/open window [1]; drag forces increase with speed [1]; larger forces from the engine are needed to overcome drag [1]; the engine provides a larger force when a car accelerates [1]; fuel provides energy for the engine to provide a larger force [1].

Force and acceleration

1a As the force increases, the acceleration increases. [2]

b Yes, they have [1], you can see a clear pattern [1] OR No, they haven't [1], you can only say that the pattern applies up to forces of 5 N [1].

c To improve the reliability of the data by identifying anomalous results [1], allowing random errors to be identified [1].

2a F = m × a [1] = 4 × 5 = 20 [1] N [1]

b Measure the size of the force applied to the rock using a forcemeter [1]. Measure its acceleration using the accelerometer [1]. Mass = force/acceleration [1]

Pages 84–85

Balanced forces

1a i For example: gravity/reaction forces [2]

ii tension in rope/force exerted by the person on the rope [2]

b When the team moves, the weight/gravity stays fixed [1], and the forces exerted by the right-hand team become larger than the forces exerted by the left-hand team [1].

2 Any 5 from: the weight of the person stays constant [1] the force from the lift is greater than weight as the lift accelerates upwards [1] it equals weight as the lift travels at a steady speed [1] it is less than the weight as it slows down [1] forces are balanced when the lift travels at a steady speed /when it is stopped [1] direction of the resultant force is upwards at the start of the trip [1] direction of the resultant force is downwards at the end of the trip [1]

Stop!

1a thinking distance [1], braking distance [1]

b less experienced/may not recognise hazards [1]

c i thinking distance increases with speed; or thinking distance doubles as speed doubles, etc [2]

ii 9 m [1]

iii The thinking distance is 6 m with no drink [1], so thinking distance has increased [1].

2 Credit valid points: 1 mark per example (maximum 2); plus 1 mark per explanation (maximum 2); plus 2 marks for choosing a factor and explaining why it has most effect.

For example: worn tyres [1] have less grip/less friction on road/more likely to skid [1];

worn brake pads [1] reduce stopping force applied to wheels [1]; worn windscreen wipers or dirty windscreen [1] reduces visibility increasing thinking distance [1].

Evaluation factor, for example: the condition of tyres is most important, as these are the only point of contact a car has with the road [2] OR the condition of the brake pads as these control the force applied to the wheels [2].

Terminal velocity

1a Terminal velocity is the top speed reached by a moving object. [1]

b i Weight is constant and greater than drag forces, so the resultant force is down [2].

ii Drag forces increase to match weight, which does not change. At terminal velocity, there is no resultant force. [2]

2a Weight remains constant [1]; drag forces increase with speed and surface area [1]; opening the parachute increases the surface area [1]; the drag forces at that speed are greater than weight [1]; the resultant force is upwards which causes deceleration [1].

b The skydiver can only hover if upward forces match weight, i.e. there is no resultant force [1]; when they are not moving up or down [1] there is a drag force upwards only if skydiver is falling [1]. A very large surface area reduces the terminal velocity greatly but does not eliminate it [1]. (Note: credit correct and relevant discussion, for example upward thermal forces may match weight in some conditions.)

Forces and elasticity

1a independent variable – force applied [1]

b dependent variable – extension of spring [1]

c He may make a mistake in his calculations/it is harder to compare his data with the other groups [1]

d He must control other variables, e.g. the spring used/ how measurements are taken [1] so only one factor is changed at a time [1].

2 (Credit correct responses – the test should relate to the quality being tested.)

Suitable qualities include: how strong or elastic the rubber is; how consistent the batches of rubber are; whether it snaps easily. [3]

Tests include: making catapults in an identical way using samples from different batches and using them to fire projectiles to see if the projectile travels the same distance each time; testing catapults to destruction. [2]

Pages 86–87

Energy to move

1a chemical [1] heat [1] sound [1]

b The rate of energy transferred depends on how quickly work is done against friction [1], friction is less on a smooth road [1].

2 Credit correct points – relevant factors include: the proportion of energy transferred to the flywheel [1]; the time the energy can be stored [1]; how the energy is transferred from the flywheel [1]; physical size [1]. (maximum 2)

An explanation of the impact of each factor should be included, for example: more energy transferred to the wheel means less transferred to electricity production [1]; the flywheel may store energy for minutes/hours, but not for days [1]. (maximum 2)

A statement of which factor is considered most important with some explanation. [1]

Working hard

1a 12 J [1]

b gravity [1] or weight [1]

c work = force x distance [1] = 2 x 25 = 50 [1] J [1]

2a 5 x 30 = 150 m [1]

b work = force x distance = 600 x 150 [1] = 90 000 J or 90 kJ [1]

c The engine is still working against friction [1], and also working against gravity; the car is gaining gravitational potential energy but has the same kinetic energy [1].

Energy in quantity

1a mass x gravity x height = 6 x 10 x 0.8 [1] = 48 [1] J [1]

b 6 x 10 x 1.4 [1] = 84 [1] J [1]

2 KE = ½ x mass x velocity² [1] = ½ x 80 x 36 [1] = 1440 J [1] for the runner

KE = ½ x 50 x 64 [1] = 1600 J [1] for the skateboarder

The child on a skateboard has more energy [1].

Energy, work and power

1a power = energy transferred/time [1] = 400 x 10 x 3/60 [1] = 200 [1] W [1]

b More power means more energy is transferred in the same time [1]; so the motor could lift a greater weight in the same time, or lift the same weight a higher distance in that time [1]. More power means the same work is done in less time [1]; so the weight is moved in a shorter time [1].

2 Approximately/the same work is done in both cases [1]. When the piano is pulled up the ramp, the force needed equals the component of its weight acting parallel to the ramp [1], so a small force acts over a large distance [1]. If the piano is lifted, the force needed equals its full weight [1], so a large force is applied over a shorter distance [1]. A force matching the weight of a piano is very hard for one person to produce [1].

Pages 88–89

Momentum

1a mass x velocity [1] = 6 [1] kg m/s [1]

b 6 kgm/s [1]

c momentum/mass [1] = 6/4 = 1.5 [1] m/s [1]

2 Conservation of momentum means momentum before and after a collision or explosion is the same provided no external factors act [1], gases are ejected from the jet packs [1], the momentum change of the moving gas in one direction equals the momentum change of the satellite in the opposite direction [1], the gas has small mass but large velocity; the satellite has a large mass and smaller velocity [1] in space, there is no air resistance so no external forces [1].

Static electricity

1a Electrons [1] are rubbed off the cloth [1] and move onto the balloon [1].

b positive [1]

c Hold the balloon near the object [1], it attracts a positively charged object [1] and repels a negatively charged object [1].

2a charge [1]

b The electroscope is given a negative charge and the gold leaf is repelled from the metal stem [1]. Then when a negatively charged object comes close, the golf leaf is repelled more [1]; when a positively charged object comes close, the gold leaf is repelled less [1].

Moving charges

1a electrostatic induction [1]

b Electrons in the wall are repelled by the negatively charged balloon [1], leaving an overall positive charge on the wall's surface [1].

2a Electric charge moves easily through electric conductors [1]; electric charge cannot move through insulators [1].

b Any three from: static electricity discharges as a spark [1]; the spark causes an electric shock [1]; at very high voltages the spark can travel through air [1]; the spark is more likely to travel to or from a point (such as a person's finger) [1].

Circuit diagrams

1a 1.6 A [1]

b 3 V [1]

c 1.5 V [1] in the series circuit, voltage is shared between components [1]

2 Any four points from: the cell transfers energy to electrons/electrical charge [1]; electrons transfer energy round the circuit/to the bulbs [1]; the bulbs transfer energy to the surroundings [1] as light/heat [1]; the total energy supplied equals the total energy used/transferred [1].

Pages 90–91
Ohm's law

1a resistance measures how easily electrons move through a material [1] OR resistance is voltage ÷ current [1]

b a number larger than 10 ohms [1]

c thickness of wire [1]; material the wire is made from [1]

2a As voltage increases, current increases [1] OR as voltage doubles, current doubles [1].

b 1.5 A [1]

c resistance = voltage/current [1] – 6/1.5 = 4 [1] ohms [1]

Non-ohmic devices

1a Change the light intensity it is exposed to. [1]

b For example: security lighting. [1]

c symbol 2 [1]

2 One mark for each example (maximum 2 marks), e.g. a thermistor's resistance falls with increasing temperature [1] an LDR's resistance falls with increases in light intensity [1] a diode has very high resistance in one direction.
One mark for each description of how the circuit responds to changes (maximum 2 marks): when the resistance falls below a certain value [1], a component (e.g. a light, a buzzer) turns on or off [1].

Components in series

1a They are all the same. [1]

b 3 V [1]

c 2 V [2]

d It increases [1]; OR it increases by a factor of 4/3 [2].

2a The potential difference of cells in series subtracts if they are connected in opposite directions [1], the total potential difference supplied is zero [1].

(**Credit** correct answers, for example the potential difference of two cells may be slightly different so there could be a small overall voltage.)

Components in parallel

1a 1.5 A [1]

b 6 V [1]

c e.g. brighter bulbs; if one bulb goes out the others stay on[1]; you can control individual bulbs[1]

2a i upper 2 ohms[1] ; **ii** lower 4 ohms [1]

b 4/3 ohms [2]

c upper branch [1] it has the lowest resistance [1]

Pages 92–93
Household electricity

1a A [1]

b The current repeatedly changes direction [1], there are 50 cycles per second [1].

2 Time for one cycle is 0.08s [1] frequency = 1/time per cycle [1] = 12.5 Hz [1]

Plugs and cables

1 Allow correct statements, any three from, for example: the wires are connected to the correct pins [1] (maximum two marks if colours and pins are correctly identified); the correct fuse is used; the outer part of the cable is gripped under the cable grip; the screws hold each wire tightly in place in each pin; check no bare wires are visible.

2 Any four from: wet skin has lower resistance than dry skin [1]; electrocution is more likely if a person has wet skin than dry skin [1]; people are more likely to have wet skin, or be sitting/standing in water in a bathroom [1]; not allowing mains sockets means that electrical equipment is less likely to be used inside the bathroom [1]; battery-operated equipment is less dangerous as it uses a lower voltage [1].

Electrical safety

1a The fuse does not cut off the electric current [1]; the cable is still live [1]; there is a risk of electrocution [1].

b Credit correct answers: the RCCB cuts off the current very quickly [1] if there is a different current in different wires in the cable [1] so there is no risk of electrocution [1].

2a The fuse can carry a current of 13 A without melting [1]. At higher currents, the fuse should melt [1].

b 5 A [1]. Current flowing in the iron is 3.47 A [1]; the fuse with the rating that is closest to the current but larger than it should be used [1].

c To protect the flex [1], from overheating/prevent fires [1].

Current charge power

1a energy = power x time [1] = 5 x 60 x 1000 [1] = 300 000 J [1] or 300 kJ

b charge = current x time [1] = 4.3 x 5 x 60 [1] = 1290 [1] C [1]

2 The potential difference [1]; doubling the potential difference doubles the energy transferred by the charge [1]. If the current is large, more energy is transferred [1], because more charge flows [1].

Pages 94–95

Structure of atoms

1a i negligible or 1/2000 [1]; ii +1 [1]; iii neutron [1].

b i number of neutrons;

  ii number of protons and electrons.

2 Any four from: electrostatic force [1]; like charged protons repel each other [1]; positively charged nucleus attracts negatively charged electrons [1]; neutrons do not feel electrostatic force [1]; strong nuclear force holds nucleons together [1].

Radioactivity

1a Radioactive – the nucleus emits ionising radiation, changing to a nucleus of a different element. [1]

b alpha [1]

c A neutron changes into a proton. [1]

2a X = 95 [1], Y = 237 [1]

b Alpha radiation cannot travel far in air/pass through skin [1]; it is only a danger if it is swallowed/inhaled/injected into the body [1].

More about nuclear radiation

1a positively charged [1]; central massive nucleus [1]; surrounded by negatively charged electrons [1]

b i deflection of some positively charged alpha particles [1]

  ii most alpha particles not deflected [1]

  iii electrons are relatively easy to remove from the atom [1]

2 Gamma radiation is not deflected as it has no charge [1]. Beta and alpha radiation are deflected in opposite directions as they have opposite charges [1]. Beta radiation is deflected more [1] than alpha radiation as it has a smaller mass [1].

Background radiation

1a 37% [1]

b Credit correct examples: some sources of background radiation increase with lifestyle choices, e.g. the number of flights taken [1]; medical history, e.g. X-rays [1]; where you live [1].

2 Credit five correct statements (for either side of the argument), for example: the Earth has always been slightly radioactive/background radioactivity is normal [1]; radioactivity can increase the lifetime cancer risk [1]; lifestyle choices/many factors affect lifetime cancer risk [1]; natural sources of background radioactivity are more significant than man-made sources [1]; there are many other significant health risks, e.g. cardiovascular disease/accidents etc [1]; lifetime cancer risk is increased in smokers by exposure to radon and thoron [1].

Pages 96–97

Half-life

1a Half-life is the time [1] taken for the original count rate/activity/mass of radioactive atoms [1] to halve [1].

b The proportions will fall [1], the carbon atoms change into nuclei of other elements [1].

2a 600 counts per second [1]

b Three [1] half-lives [1], which is 15 years [1].

Using nuclear radiation

1a For example: reduce wastage/controls the quality. [1]

b It is absorbed by cardboard. [1]

c Less radiation detected means the cardboard is too thick [1]; so the rollers should move closer together [1].

2 Pupils should cover both sides of the argument. Credit five correct statements, for example: radioactive tracers help diagnose medical conditions while avoiding surgery [1]; they can be more accurate than other methods [1]; using tracers helps people choose the most effective treatment [1]; medical tracers can increase the lifetime cancer risk [1]; medical risks are reduced by choosing suitable tracers (e.g. short half-life; not alpha-emitting) [1]; tracers should be used if more serious conditions are suspected [1].

Nuclear fission

1a When a nucleus splits into two or more products. [2]

b Nuclear power/generating electricity. [1]

c 3 [1]

d More nuclei are involved at each stage of the chain reaction [1]; because each reaction produces more neutrons than are needed to cause fission [1]; too much heat/energy will be produced if too many nuclei are involved [1].

2 Pupils should suggest at least one advantage and one disadvantage with an explanation.

Advantages – they last for a long time so reducing the need for operations to replace the battery; they are compact (so not a nuisance to the patient); they are reliable (so people will not suffer side-effects). [2]

Disadvantages – disposal of radioactive material can be difficult/expensive; patient may be exposed to unacceptable levels of radiation; potential danger if medical staff are not aware that there is a nuclear battery inside the patient. [2]

Nuclear fusion

1a In nuclear fission, nuclei split; but in nuclear fusion, nuclei join [1]; nuclear fission involves large nuclei, but nuclear fusion involves light nuclei [1].

b the Sun [1]

2 Nuclear fusion creates different elements [1]; light elements form heavier elements [1]; in stars, elements up to iron are formed [1]; supernovas generate such immense energy and heat [1] that heavier elements can form [1].

Page 98

Life cycle of stars

1 Credit correct references to changes in forces at specific stages. Any five from: gravity forces the star to collapse/increases the pressure inside it [1]; forces from heat generated in the star force it to expand [1]; during each stage the forces balance [1]; when the fuel runs out, less heat is generated and the star collapses [1]; because gravity is greater than forces from heat [1]; in very massive stars, the heat generated can be enough to cause a rapid expansion [1].

2 Any three from: more massive stars generate more heat in their core [1]; fusion reactions creating heavier elements take place in the core of heavier stars [1]; more stages in the life cycle can take place for more massive stars [1]; smaller stars may just cool and die as their fuel runs out [1]; more massive stars may explode as a supernova [1].

Page 99

Extended response question

5 or 6 marks:

A detailed description of two safety features, e.g. airbags, crumple zones, seat belts, and side impact bars. The answer should include a discussion of the momentum and energy changes taking place, linking increased impact duration with a reduced risk of serious injury (change in speed takes longer, so the deceleration and forces felt are smaller). The answer should explain how these safety measures increase the duration of the collision and/or absorb energy (e.g. seat belts stretch slightly, which extends the time a person takes to stop; crumple zones deform and absorb energy – this extends the time it takes to come to a complete halt). *All information in answer is relevant, clear, organised and presented in a structured and coherent format. Specialist terms are used appropriately. Few, if any, errors in grammar, punctuation and spelling.*

3 or 4 marks:

A limited description of two safety features e.g. airbags, crumple zones, seat belts, and side impact bars. There should be a discussion, which describes what these safety measures do, and some attempt to link this to increased impact duration or reduced energy transfer. *For the most part, the information is relevant and presented in a structured and coherent format. Specialist terms are used for the most part appropriately. There are occasional errors in grammar, punctuation and spelling.*

1 or 2 marks:

An incomplete description of one or two safety features. There should be some description of energy changes or momentum changes during a collision. *Answer may be simplistic. There may be limited use of specialist terms. Errors of grammar, punctuation and spelling prevent communication of the science.*

P3 Answers

Pages 100–101

Using X-rays

1a

Statement	Tick
X-rays cause ionisation	✓
X-rays are absorbed by bone but pass through skin	✓
X-rays have long wavelengths	
X-rays are absorbed by all tissues	
X-rays are used to kill cancer cells	✓
X-rays do not cause ionisation	
X-rays have a very high frequency	✓

b One advantage, for example: images can be viewed in real-time [1] which speeds up treatment [1].

2 One mark for each precaution (max 2 marks), e.g. wear protective clothing; stand behind screen; do not stay in room with patient.
One mark for each explanation (max 2 marks), e.g. reduces exposure by reducing intensity/reduces time of exposure.

Ultrasound

1a One mark for each correct (bold) word.
X-rays are **absorbed** by bone and **transmitted** through soft tissue.
This creates a **shadow** picture that can be used for a medical diagnosis.
Ultrasound is partly **transmitted** and partly **reflected** at a boundary. [either order gains the marks]
Reflections are **analysed** at a detector to create an image.

b Ultrasound is non-ionising [1]. It does not damage DNA in cells [1]; a fetus' cells are dividing and are vulnerable to damage caused by ionisation [1].

2a One peak is the original pulse [1], one peak is the reflected pulse [1].

b The reflected peak [1] has been partially absorbed [1].

c distance there and back = speed × time [1 mark] = 1500 × (0.02 × 5) /1000 [1]. Eyeball size is half this distance = 7.5 [1] cm [1]

Refraction

1a power = 1/f [1] = 1/0.1 = 10 dioptre [1].

b The focal length is shorter [1]; more powerful lenses bend light more [1].

2 If two lenses have the same focal length, increasing the refractive index of the material [1] means the lens can be flatter [1].
The optician should make the stronger lens out of material [1] that has a larger refractive index [1].

Pages 102–103

Lenses

1a magnification is image height/object height [1] = 5.5/1 = 5.5 [1 mark]

b The tree will be taller [1].

2a The image is inverted [1], diminished [1], real [1].

b The image is upright [1], magnified [1], virtual [1].

Seeing clearly

1a retina [1]

b Ciliary muscles contract or relax [1] so suspensory ligaments are tense or loose [1] which flattens the lens or makes it fatter [1].

2a short sight

b He can use a concave lens [1] to increase the focal length of his eye's lens [1] so light is focused on the retina [1].

More uses of light

1a The ray undergoes total internal reflection/reflects repeatedly [1] off the inner sides of the optic fibre [1].

b

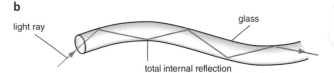

light ray

glass

total internal reflection

2 refractive index = 1/sin c [1] = 1/sin 41° [1] = 1.5 [1]

Centre of mass

1 Suspend the tennis racket from at least two places [one at a time] [1], draw a line running vertically down from each point of suspension [1]. The centre of mass is where the lines cross [1].

2 The lorry topples if its centre of mass does not lie over its base [1], so the centre of mass should be as low as possible [1]. Heavy goods should be loaded first [1] and spread out evenly across the lorry's trailer [1].

Pages 104–105

Moments

1a 30 × 0.7 [1] = 21 Nm [1]

b 21 Nm

c 21/0.5 [1] = 42 N [1]

2a 60 000 × 10 [1] = 600 000 N [1]

b 600 000/30 [1] = 20 000 N [1]

c Moments clockwise and anticlockwise must match [1] the counterbalance should be heavier [1] to increase the clockwise moment [1] or the hooks should move closer to the pivot [1] to reduce the anticlockwise moment [1].

Hydraulics

1a P = F/A [1] = 25/0.001 [1] = 25 000 Pa [1]

b F = PA = 25 000 × 0.005 [1] = 125 N [1]

2a It multiplies forces [1] so the small force from the brake pedal becomes a large force at the wheels [1]; it transfers forces round corners [1] so pressing the brake pedal applies a force at all the wheels [1].

b Brake fluid should not freeze or evaporate in the temperature range that the car is likely to be used in [1] as the force isn't transmitted through gas or solids [1]. The brake fluid should not expand and contract as the temperature changes [1], otherwise the same force isn't transmitted at different temperatures [1].

Circular motion

1a centripetal force

b i the force increases in size [1] the direction does not change [1].

ii the force increases in size [1] the direction does not change [1].

2 Acceleration is change of velocity (speed with direction) [1]. The speed is fixed but the direction changes [1].

Circular motion in action

1a 2 seconds

b frequency = 1/time period [1] = ½ [1] = 0.5 Hz [1]

c It does not change [1], the time period depends on the length of the pendulum only.

2 At the top of the swing, the ride does not move/has maximum gravitational potential energy [1]. At the bottom of the swing, it has maximum kinetic energy/moves at its fastest [1]. The total energy does not change through the ride [1].

Pages 106–107

Electromagnetic force

1   1 = C
       2 = D
       3 = B
       4 = A   [4]

2 The motor effect is the a force created when a current flows perpendicular to a magnetic field [1]. In the drill, a coil of wire can rotate between fixed magnets/in a magnetic field [1].
When the drill is turned on, a current flows in the coil, and it spins [1].
The drill bit is attached to the coil and also spins [1].

Electromagnetic induction

1a It goes to zero [1].

b It gives a negative reading [1], the same value as when she moved the coil in [1].

c Use a coil of wire with more turns [1]. Move the coil faster [1].

2 When equipment is turned on, the current changes [1]. This induces a second current to flow inside the equipment [1]. The additional current may be enough to melt the fuse [1].

Transformers

1a It's a step-up transformer [1] because there are more turns on the secondary coil [1].

b The current must flow through the coils of wire, not the metal core [1].

c A current is only induced in the secondary coil if the magnetic field is changing [1], the changing current induces a changing magnetic field [1].

2 turns on primary/turns on secondary = voltage on primary/voltage on secondary [1]
output voltage = 12 × 25 000/1000 [1] = 300 [1] V [1]

Using transformers

1a No energy is lost during energy transfers in the transformer. [1]

b power = current × voltage [1] = 1.6 × 400 [1] = 640 V [1]

c 640 V [1]

d current = power/voltage = 640/120 [1 mark] = 5.33 A [1]

2 State advantages: e.g. switch-mode transformers are smaller, lighter and use less energy when equipment is not being used [1]; using switch-mode transformers saves energy and reduces electricity bills [1].
State disadvantages: e.g. it will cost money to install them in existing equipment [1].
State a conclusion with a reason: e.g. they should be fitted if the savings are greater than the installation cost [1], and if it is possible to install them [1].

Page 108
Extended response question

5–6 marks

A detailed description of relevant properties of X-rays and ultrasound should be included (e.g. ionising ability, whether they are reflected or absorbed by different tissues). The medical use and limitations of each type of wave should be linked to its properties, e.g. rapid vibrations from ultrasound waves cause kidney stones to break up.

All information in answer is relevant, clear, organised and presented in a structured and coherent format. Specialist terms are used appropriately. Few, if any, errors in grammar, punctuation and spelling.

3–4 marks

The answer will include a limited description of the relevant properties (including ionising ability) of X-rays and ultrasound. At least one medical scan or treatment for each type of wave should be stated but may not be linked to their properties. Pupils should also mention limitations for each type of wave.

For the most part the information is relevant and presented in a structured and coherent format. Specialist terms are used for the most part appropriately. There are occasional errors in grammar, punctuation and spelling.

1–2 marks

The answer includes an incomplete description of relevant properties of X-rays and/or ultrasound. These may be linked to specific uses. At least one relevant use of X-rays and/or ultrasound should be stated. Pupils may also mention some limitations.

Answer may be simplistic. There may be limited use of specialist terms. Errors of grammar, punctuation and spelling prevent communication of the science.